ENVIRONMENTAL CONNECTIONS

A Guide to Environmental Studies

Kathleen B. deBettencourt

Matthew Feeney

A. Nicole Barone

Keith White

KENDALL/HUNT PUBLISHING COMPANY
4050 Westmark Drive Dubuque, Iowa 52002

The Environmental Literacy Council is a non-profit organization established to help educators learn and teach about often complex, interdisciplinary environmental issues. The Council brings together scientists, economists, educators, and other experts to enrich environmental studies.

Environmental Literacy Council
1730 K Street NW, Suite 905
Washington, DC 20006
202-296-0390
www.enviroliteracy.org

ISBN 0-7872-7105-5

Printed in the United States of America.
10 9 8 7 6 5 4 3 2 1

"You see," Father Worm began, "Harriet loved Nature. But loving Nature is not the same as understanding it. And Harriet not only misunderstood the things she saw—vilifying some creatures while romanticizing others—but also her own connection to them." Father Worm paused, his eyes narrowing. "Ah, connections, Son. That's the fateful key that Harriet missed, the key to understanding the natural world."

★Gary Larson (creator of the Far Side), There's a Hair in my Dirt!★

We dedicate this to all those who have the difficult task of helping young minds make connections between what is contained in the pages of a textbook and the world outside the classroom. We thank all those who offered comments, suggestions of books and resources, and particularly members of the Environmental Literacy Council who generously contributed their time and expertise to this project.

CONTENTS

PURPOSE

The purpose of this guide is to assist teachers who teach an environmental science course or who include environmental topics in their curriculum. Environmental science is an interdisciplinary subject that draws from biology, chemistry, physics, economics, history, and other disciplines. It is also an area in which scientific knowledge is rapidly growing. Many textbooks include a chapter or some discussion of environmental topics, but often the presentations are cursory, out of date, or provide inadequate coverage of important scientific concepts that would help the reader to understand the issue. This guide, therefore, is intended to help educators and students find resources that would enable them to examine these issues in greater depth. We also hope to raise curiosity that might inspire at least some readers to study the environmental sciences further.

We have provided a brief summary of a number of environmental issues that are often mentioned in textbooks, in the media, and in scientific journals. For each, we have tried to locate a range of resources that would be helpful in researching the topic, including those textbooks that cover the scientific concepts in some depth, online and print articles that explain the topic in clear and simple terms, sources of data for further analysis, and books and articles from scientific journals that present the current state of scientific understanding. Environmental issues rarely present questions that can be resolved by science or technology alone; therefore, we have included resources that analyze the economic and policy aspects of issues and address the risks and tradeoffs inherent in decisionmaking.

We have not attempted to cover every environmental issue. In any case, new issues appear on the pages of newspapers and journals almost daily and new evidence from ongoing scientific research often raises new questions. Nor is this intended to be an exhaustive list of resources that should be considered in researching any particular question. This is, rather, intended to be a beginning. There is a rich and growing variety of books, journals, and Internet sites devoted to the environmental sciences. We hope to familiarize the reader with the resources, the online databases, the scientific journals, and other sources of information that are available for research. This guide is designed to be used in conjunction with our web site (http://www.enviroliteracy.org), which is periodically updated to include new resources and articles as they become available.

The pace of development of new technologies in medicine, communications, energy, and manufacturing was more rapid in the twentieth century than ever before in human history and has resulted in profound changes in social mobility, education, and health. At the turn of the twentieth century, most of the world's population lived in rural areas. Medical technology was limited and many people died at a young age in childbirth, epidemics, or from respiratory illnesses. Few homes had electricity or telephones. Transportation, either by horse or railroads, was slow and difficult. In the few decades since the end of World War II, vast improvements in health, comfort, and convenience have been realized by developed nations, although meeting basic human needs remains a formidable challenge for a substantial portion of the world's population.

There have also been enormous advances in our ability to understand the natural world. New instruments, such as computers, satellite imagery, electron microscopes, and mass spectrometers, have permitted us to observe, measure, and examine parts of the Earth that were previously inaccessible. These new tools have deepened our understanding of the intricate inter-workings of Earth's systems of air, water, earth, and living things. They have also led to new understanding of the impact of human activities on the environment.

Humans have modified their environments since they first cleared land for villages and planted crops. When populations were small and dispersed, however, the effect on the environment was local and limited. A larger population and an expanded agricultural and industrial capacity have increased human use of natural resources. The challenge of the new century is to preserve the quality of life that we have achieved and to meet the basic human needs of a growing population in ways that minimize costs to natural resources and living things. Demands for water, land, mineral resources, and energy will continue to grow, and production of the goods and services that have transformed health and welfare for human populations will require natural resources.

This is not a small challenge, and for the foreseeable future there will continue to be uncertainties and disagreements about the most appropriate policies and actions. Issues relating to natural resources and the environment will be at the forefront of the public agenda for decades to come. The quality of the decisions that we make about these issues depends on how knowledgeable citizens, policy makers, and business leaders are about the relationships between humans and the natural world and the interactions among Earth's systems. There are always costs and tradeoffs in choosing one policy over another, and often there are unanticipated, unintended consequences. A deepened understanding of these tradeoffs will permit a prudent reckoning of what is to be gained and what may be lost.

This guide is intended to provide a basic introduction to the major issues relating to natural resources and environmental quality that have been and will be on the public agenda as we begin the 21st century. These are all issues of some importance that may have potential impacts on health and quality of life. These are issues about which reasonable people will continue to disagree, in which competing interests and values will be put forth, and which, in a democratic society, must ultimately be decided by citizens. This guide is intended to assist with unraveling some of the complexities of these issues with the hope of fostering understanding and thoughtful decision making.

Each section includes an overview of the issue, summarizing what is known and not known, the risks and tradeoffs of proposed actions, and the significant areas of disagreement. For each issue, annotated resources are suggested to provide background information. Sources of data for further research and analysis are included, in addition to references to significant legislation, regulations, and policies related to the topic. For each topic, there are resources included that are readily accessible and written to explain the issue in simple terms to a layperson. For further investigation of the topic, we have also noted resources from scientific journals and books written for audiences with some expertise.

We have, in the necessity of imposing some organizational structure, separated issues into sections such as air, water, and soils. Environmental concerns are, however, "profoundly systemic." The environmental sciences study the processes by which matter and energy flow through the Earth's systems of air, water, land, and living things, and analyze how human actions affect those processes. The planet is a complex and dynamic system whose liquid, gaseous, solid, and organic realms are closely linked, and a change in one may lead to a change in all the others. It is often misleading to think of one issue in isolation from another. The question of which energy sources are to be developed, for example, cannot be separated from issues relating to climate, water, or air quality. Changes in land use, such as clearing land for agriculture, have impacts on climate, soils, water quality, and biodiversity, among other impacts. We have tried to indicate these complex interactions in each section.

How To Find Resources

Environmental issues are complex because they incorporate scientific knowledge from the natural and social sciences and entail social, economic, historical, and ethical considerations. For background on these issues, the following resources may be helpful:

Textbooks

For the basic scientific concepts relating to environmental topics, introductory college-level textbooks in the Earth sciences, chemistry, biology, ecology, or physics are useful. Some of these textbooks provide introductions to concepts (such as the pH scale) that are necessary for discussions of environmental issues. Textbooks can be expensive, but they can often be found in school, public, or university libraries.

Books

There are a number of basic books written to explain environmental issues to the layperson, and we have made suggestions of several. Many of these are available in a well-stocked public library or they can be ordered from a bookstore or bookseller through the Internet.

Articles

Scientific journals, such as *Science* and *New Scientist*, and magazines, such as *Scientific American* and *Discover*, provide clearly written discussions of environmental topics. Many of these are available in school, public, or university libraries. Many of these journals make full-text articles available on their Internet sites.

Electronic databases

Most public (and many school) libraries now have several online article databases that allow the investigator to search for article references and often retrieve the full article text. General databases are common, but a variety of subject-specific databases are also available in topics such as Science, Economics, Business, or Literature. These databases can be searched by title, author, subject, keyword, or a combination of the above. The most useful general article database that we have found is *Periodical Abstracts*, which includes articles from 1979 to the present from national weekly periodicals and scientific journals.

Bibliographies

Most textbooks and books include footnotes and bibliographies of books and articles that provide further information on a topic.

Internet

The Internet developed as researchers in a number of universities around the world began to establish computer-based networks to facilitate sharing of scientific research and data. It has evolved beyond its original mission, but it remains an amazing achievement in providing access to information and data.

GUIDELINES FOR FINDING AND EVALUATING INFORMATION

Knowledge is of two kinds. We know a subject ourselves, or we know where we can find information upon it.

★Samuel Johson★

The Internet makes information available faster than ever before, unfiltered and immediate. This means, however, that it is important to evaluate the credibility and accuracy of information. Anyone with access to a computer and a telephone can publish on the Internet. How can you tell if a web site is a credible source of information? Here are some things to look for:

- Does the author of the information have academic credentials in the field in which he or she is writing? The web site should include information about the author's background and academic degrees. If no author is listed, that is a warning sign.
- Has the article been published in a peer-reviewed journal, such as *Science* or the *New England Journal of Medicine*? Publication in a peer-reviewed journal means that other scientists in the field have made a judgment that the author's work is credible. Other publications, such as *Scientific American*, while they are not peer-reviewed journals, have a reputation of careful writing on science issues.
- Does the author cite the sources of data upon which he or she is relying? Does the author explain how and when the data were collected?

Some sites to check:

- Good sources of information can generally be found at universities and research institutions, such as Scripps Oceanographic Institute or Lawrence Livermore National Laboratory, where there are experts engaged in ongoing research.

- Government agencies such as NASA and the Department of Energy have the resources to make a vast array of information and educational resources available to researchers, teachers, and students.
- Several sites specialize in providing pro and con discussions of environmental issues, such as *http://www.policy.com* and *http://www.intellectualcapital.com*. On each site, journalists, policy analysts, and scientists post articles that argue various positions on matters of public policy.
- The National Library for the Environment compiles articles prepared by the Congressional Research Service (CRS). These articles cover all major areas of environmental policy and range from basic explanations to in-depth analyses of the economic implications of policy options. The CRS provides these articles for use by members of Congress and their staffs, so they are easy to understand. *http://www.cnie.org/nle*
- *New Scientist* magazine's online site includes coverage of current scientific discoveries and controversies, with a bit of British humor. Feature articles present the uncertainties surrounding issues such as climate change and genetic modification. Search the archives for past issues or search by topic. *http://www.newscientist.com*

General Reference Books

The René Dubos Center for Human Environments. *The Encyclopedia of the Environment*. Boston: Houghton Mifflin, 1994.
This reference book includes brief overviews of fundamental environmental issues and concepts written by over three hundred scientists, economists, and other experts. This book is out of print, but it should be available in a public library.

Kemp, David. *The Environment Dictionary*. New York: Routledge, 1998.
This is a basic resource that includes detailed definitions of environmental terms and concepts, including chemical formulas, data tables, maps, and illustrations.

Nierenberg, William, ed. *Encyclopedia of Environmental Biology*. New York: Academic Press, 1995.
This three-volume encyclopedia contains articles explaining major environmental biology concepts and processes by experts in the field.

World Resources Institute. *World Resources 1998-99: A Guide to the Global Environment*. New York: Oxford University Press, 1998.
A joint effort of WRI, the World Bank, and the United Nations, this guide includes brief overviews of global trends in population, food production, deforestation, and other environmental issues, with extensive data. Information on how to order is available at *http://www.wri.org* or at 1-800-822-0504. A new edition was to be available in mid-2000.

Lemons, John, ed. *Scientific Understanding and Environmental Problem Solving*. New York: Blackwell Science, 1995.
This collection of articles assesses current scientific understanding of issues such as water quality, ecosystem health, nuclear waste, and fisheries management and addresses the implications of environmental decision making in the face of scientific uncertainties. This is an expensive volume; it may be available as a reference book in public or university libraries.

United Nations Environment Programme. ***Global Environment Outlook-1: Global State of the Environment 1997.*** **Cary, NC: Oxford University Press, 1997.**
This report from the United Nations Environment Programme assesses global trends in environmental issues. Detailed indicators of environmental quality are included by country and region and the most significant environmental problems for each region are summarized. A web version is also available at *http://www.unep.org/unep/eia/geo1/index.htm*

General Reading

Golley, Frank. ***A Primer for Environmental Literacy.*** **New Haven: Yale University Press, 1998.**
This book is intended to introduce readers to the energy dynamics of Earth's systems and its interacting subsystems: the atmosphere, hydrosphere, lithosphere, and biosphere.

Pickering, Kevin, and Lewis Owen. ***An Introduction to Global Environmental Issues.*** **New York: Routledge, 1994.**
This college-level text provides more detail about the scientific aspects of environmental issues than typical introductory-level texts.

Manaham, Stanley. ***Environmental Science and Technology.*** **New York: Lewis Publishers, 1997.**
This college textbook provides detailed explanations of the fundamentals of chemistry, biology, biochemistry, and environmental chemistry. This is for advanced students, but the text is a useful reference book for chemical formulas and other information not generally included in introductory environmental science texts.

Wallace E. Oates, ed. Resources for the Future. ***The RFF Reader in Environmental Resource Management.*** **Washington, DC: Resources for the Future, 1999.**
Over forty articles from *Resources* magazine have been bound into this single volume. Topics include climate change, forests, water, environmental equity, environmental policy, risk, and economics. Available for purchase at *http://www.rff.org/* or by calling 1-410-516-6955.

Bailey, Ronald, ed. ***Earth Report 2000: Revisiting the True State of the Planet.*** **Washington, DC: McGraw-Hill, 2000.**
This collection of articles takes an optimistic view that human ingenuity and technology will provide answers to environmental problems. Ordering information is available at *http://www.cei.org* or by calling 1-202-331-1010.

Piel, Jonathan, ed. ***Managing Planet Earth: Readings from Scientific American.*** **New York: W. H. Freeman and Company, 1989.**
This is a collection of articles from *Scientific American* that provide basic overviews of major environmental issues and strategies to address them. Although published in 1989, many of the articles are still relevant.

Goudie, Andrew. ***The Human Impact on the Natural Environment.*** **Cambridge, MA: MIT Press, 1996.**
This is an extensive examination of the impacts humans have had on the environment since pre-recorded times. It is rich in historical detail and provides a fairly comprehensive overview of the scientific literature in this area.

National Research Council. ***Our Common Journey: A Transition toward Sustainability.*** **Washington, DC: National Academy Press, 1999.**
This report presents a thorough review of the most pressing environmental concerns worldwide, assesses the current state of scientific knowledge in each area, and makes recommendations for a research agenda that would provide the knowledge necessary to support sustainable development. The report can be ordered or read at *http://www.nap.edu*

Hammond, Allen. ***Which World? Scenarios for the 21st Century.*** **Washington, DC: Island Press, 1998.**
Considering advances in technology, a growing population, and finite natural resources, World Resources Institute senior scientist Hammond reviews three possible options for the future: a market-driven world, a collectivist world, and a catastrophic world.

John McPhee. ***Encounters with the Archdruid.*** **New York: Farrar, Straus, and Giroux, 1971.**
The nuances of differences in environmental perspectives are explored in McPhee's accounts of separate "managed confrontations" between environmentalist David Brower and three noted natural resource developers: mineral engineer Charles Park, resort designer Charles Fraser, and dam builder Floyd Dominy.

McPhee, John. ***The Control of Nature.*** **New York: Farrar, Straus, and Giroux, 1989.**
Just how effectively can humans control nature? McPhee describes some successes, but notes that final victory is at best elusive, and possibly unattainable.

McClellan, James E., and Harold Dorn. ***Science and Technology in World History.*** **Baltimore: Johns Hopkins University Press, 1999.**
The authors review the history of science and technology from the Paleolithic era, when humans first began to use tools and wonder about nature, to contemporary times. They demonstrate how development of various technologies, such as metal plows, yokes for draft animals, and the steam engine, affected human existence.

Bowler, Peter J. ***The Environmental Sciences.*** **New York: W.W. Norton & Co., 1992.**
This book, from the Norton History of Science series, chronicles the origins of fields, such as Earth science, ecology, geophysics, and geology, which are now considered part of the environmental sciences.

Gregg Easterbrook. ***A Moment on the Earth: The Coming Age of Environmental Optimism.*** **New York: Viking Press, 1995.**
This book by journalist Easterbook is intended to offer a more optimistic assessment of the state of the environment than is often seen in the media.

Lewis, Martin W. ***Green Delusions: An Environmentalist Critique of Radical Environmentalism.*** **Durham, NC: Duke University Press, 1992.**
Lewis, a self-described environmentalist, counters those who believe in a get-back-to-nature ideal. Lewis reviews evidence that suggests primitive societies were not necessarily peaceful or environmentally benign and argues that technology and economic growth can be beneficial.

Ausbel, Jesse H., and H. Dale Langford. ***Technological Trajectories and the Human Environment.*** **Washington, DC: National Academy Press, 1997.**
This collection, originally published in the journal *Daedulus* [Vol. 125 (3), 1996], addresses the interrelationship between technology and the environment and includes articles that review the progress made and promise of technological evolution in agriculture, energy, and other fields of importance to the environment. Available at *http://www.nap.edu*

Ausbel, Jesse H., David G. Victor, and Iddo K. Wernik. "The Environment Since 1970." ***Consequences,*** **Fall 1995.**
This paper reviews data available from government sources to "stake stock" of environmental change in the twenty-five years since the first Earth Day. The authors assess trends in population, energy, health, agriculture, and environmental quality. Available at *http://www.gcrio.org/CONSEQUENCES/fall95/envir.html*

LABORATORY MANUALS AND FIELD GUIDES

There are a number of lab manuals and activity guides available for the environmental sciences. Each has some useful material; several include quantitative exercises that would be helpful. Following are a few suggestions. In addition, the Environmental Literacy Council web site (*http://www.enviroliteracy.org*) has a lab exchange for teachers that includes interesting exercises on a number of topics.

Rosenthal, Dorothy. ***Environmental Science Activities.*** **New York: John Wiley & Sons, Inc., 1995.**
This book includes forty-six simple activities and demonstrations, most of which are conducted outdoors and use readily available materials. Activities include marking and capturing pillbugs to estimate population size, using bioindicators to monitor environmental pollutants, and calculating the Earth's carbon budget.

Cothron, Julia, Ronald Giese, and Richard Rezba. ***Students and Research: Practical Strategies for Science Classrooms and Competitions.*** **Second Edition. Dubuque, IA: Kendall/Hunt Publishing Co., 1993.**
This manual is intended to assist teachers in educating their students about basic research methods. It reviews, among other things, how to design experiments, analyze data, determine statistical significance, and report findings.

University of Wisconsin, Madison. ***Bottle Biology.*** **Dubuque, IA: Kendall/Hunt Publishing Co., 1993.**
This National Science Foundation-funded project uses plastic soda bottles for activities that demonstrate ecosystem concepts such as decomposition, predator/prey relationships, and plant growth.

McConnell, Robert, and Daniel Abel. ***Environmental Issues: Measuring, Analyzing, and Evaluating.*** **Upper Saddle River, NJ: Prentice Hall, 1999.**
This college-level environmental science workbook demonstrates how to quantitatively analyze environmental issues. Using questions related to issues such as energy, climate, and population, the authors provide detailed instructions that show how to determine

what data would be necessary to answer the question, how to find the data, and how to analyze it.

Enger, Eldon, and Bradley Smith. *Field and Laboratory Activities in Environmental Science*. Sixth Edition. Boston: WCB/McGraw-Hill, 1997.
This lab manual accompanies an introductory college level environmental science textbook by the same authors, and includes basic activities and experiments, including measuring dissolved oxygen and testing pH levels in water samples. It also includes exercises that review chemistry basics, including acids, bases, and chemical reactions.

Lawrence, David. *Earth Science: A Graphing Calculator Approach*. Fairfield, NJ: Pencil Point Press, Inc., 1999.
This manual demonstrates using a graphing calculator for environmental science-related exercises, including making a climatograph, statistically analyzing temperature variation, and simulating a map of the ocean floor.

Trautmann, Nancy, and Marianne Krasny. *Composting in the Classroom: Scientific Inquiry for High School Students*. Dubuque, IA: Kendall/Hunt Publishing Co., 1998.
The manual gives an overview of composting science; provides instructions on how to build, mix the ingredients, and measure the attributes of the compost; and how to design meaningful research projects involving the compost.

Johnson, Robyn, Scott Holman, and Dan Holmquist. *Water Quality with Computers*. Portland, OR: Vernier Software, 1999.
This manual provides instructions for sixteen water quality tests that can be conducted using Vernier sensors with *Logger Pro*. A CD-ROM with experiment files is included with manual.

Johnson, Robyn, Scott Holman, and Dan Holmquist. *Water Quality with CBL*. Portland, OR: Vernier Software, 1999.
This book provides instructions for sixteen water tests that can be done using Vernier sensors with the Texas Instruments Calculator Based Lab (CBL) System and graphing calculators. A CD-ROM with experiment files is included with manual.

KrumHardt, Barbara, and Danielle Wirth. *Experiences in Environmental Science*. East-Dubuque, IL: Bellwether-Cross Publishing, 1999.
Intended for non-science majors at the college level, this manual provides environmental science labs covering populations, agriculture, water, biodiversity, and waste. The book includes some field activities not included in typical environmental science lab manuals, such as an exercise in which students estimate the commercial value of timber sawlogs, a study of demographics using a local cemetery, and writing an environmental impact statement.

Arnfield, Edwin, Angela Bitterman-Wenson, and Alka Ahuja. *Environmental Science: A Field and Laboratory Manual*. Revised Printing. Dubuque, IA: Kendall/Hunt Publishing Co., 1999.
This lab manual provides instructions for conducting basic environmental science exercises including estimating population densities, modeling waste water treatment, analyzing roadside and plants, and calculating the efficiencies of various insulating materials.

A Historical Perspective

Since the beginning of recorded history, people have wondered about the relationship of humans to the natural world. Herodotus, generally acknowledged to be the first known historian, described in detail how the manners, culture, and political organization of ancient civilizations such as the Egyptians, Babylonians, and Libyans were shaped by the characteristics of the land in which they settled. He also noted the overgrazing that had occurred in lands surrounding the Mediterranean. In the *Critias*, Plato described the severe erosion caused by deforestation in Attica. When survival was the paramount consideration for the majority of human populations, however, few gave thought to the consequences of human activity on the land, and indeed, when population numbers were small, the impacts were largely local and limited.

> In the past the Grand Canyon invited reflection on human significance, but today much of the public sees it through a cultural lens shaped by advanced technology. The characteristic questions about the canyon reported by Park Service employees assume that humans dug the canyon or that they could improve it so that it might be viewed quickly and easily. Rangers reported repeated inquiries for directions to the road, the elevator, the train, the bus, or the trolley to the bottom. Other visitors request that the canyon be lighted at night. Many assume that the canyon was produced either by one of the New Deal dam-building programs or by the Indians–"What tools did they use?" is a common question.
>
> ★David E. Nye, 1995, as quoted in **Visions of Technology: A Century of Vital Debate about Machines, Systems, and the Human World★**

When the Europeans first settled in the Americas, they regarded it as a vast expanse of seemingly limitless natural resources. This view was expressed by Hector St. John de Crevecoeur in his *Letters from an American Farmer,* written in 1782:

> Many ages will not see the shores of our great lakes replenished with inland nations, nor the unknown bounds of North America completely peopled. Who can tell how far it extends? Who can tell the millions of men whom it will feed and contain?

From early colonial times some individuals raised attention to wasteful land management practices by colonial farmers. Swedish scientist Peter Kalm, for example, visited America to identify plants and collect seeds that might thrive and be useful in the environment of his homeland. He recorded his observations of the American colonies in *Travels in North America*, a witty, intelligent account describing in detail his three-year trek (1748-1752). Kalm reported that:

> The rye grows very poorly in most of the fields, chiefly because of careless agricultural practices and the poor soil, which is seldom or never manured. After the inhabitants have converted into a tillable field a tract of land which was forest for many centuries, and which consequently had a very fine soil, they use it as long as it will bear any crops. When it ceases to bear any, they turn it into pastures for cattle and take new grain fields in another place, where a rich black soil can be found that has never been used. This kind of agriculture will do for a time, but it will afterwards have bad consequences.

The American founders were also concerned that the abundance of land in America had led to wasteful practices. James Madison, for example, wrote in 1818, "With so many consumers of the fertility of the Earth, and so little attention to the means of repairing the ravages, no one can be surprised at the impoverished face of the country."[1] Thomas Jefferson was as interested in natural science as he was in human things, and his *Notes on the State of Vir-*

ginia (1782), written in response to a Frenchman's inquiry, is a detailed exploration of the natural environment of the state as well as its culture and manners. He says of its farmers:

> The indifferent state of agriculture among us does not proceed from a want of knowledge merely; it is from our having such quantities of land to waste as we please. In Europe the object is to make the most of their land, labor being abundant; here it is to make the most of our labor, land being abundant...

Another visitor to America, Frenchman Alexis deTocqueville, in his 1831 *Democracy in America*, described the abundance of land in America and its effect on its people:

> In Europe people talk a great deal about the wilds of America, but the Americans themselves never think about them; they are insensible to the wonders of inanimate nature and they may be said not to perceive the mighty forests that surround them till they fall beneath the hatchet. Their eyes are fixed upon another sight: the American people views its own march across these wilds, draining swamps, turning the course of rivers, peopling solitudes, and subduing nature.

The Romantic Movement: Emerson and Thoreau

A romantic, spiritual view of the natural world was popularized by a small group of New England intellectuals, the transcendentalists. The acknowledged leader of the transcendental thinkers was Ralph Waldo Emerson (1803-1882), a former Unitarian minister turned secular preacher, platform lecturer, essayist, and poet. In Emerson's first book, *Nature*, he speaks of finding beauty and wisdom in the natural world:

> In the woods too, a man casts off his years, as the snake his slough, and at what period so ever of life, is always a child. In the woods, is perpetual youth... In the woods, we return to reason and faith.

Emerson's younger colleague Henry David Thoreau is best remembered as the author of *Walden, or, a Life in the Woods* (1854) which was described by historian Arthur Ekirch as "the first real attempt by an American to work out a philosophy of man and nature." *Walden* is an account of Thoreau's experience living for twenty-six months in a shack he built on Emerson's property at Walden Pond, about a mile from Thoreau's home in the village of Concord. Thoreau's motive was to conduct an experiment in simple living: "I went to the woods because I wished to live deliberately, to confront only the essential facts of life, and see if I could not learn what it had to teach, and not, when I came to die, discover that I had not lived." Thoreau and Emerson wrote in poetic language of the need for "wildness" as offering relief from the labors, crowds, and materialism of urban civilization.

The Emergence of Conservation: George Perkins Marsh

Possibly the first detailed analysis of the impacts of human activities on the environment was George Perkins Marsh's *Man and Nature, or, The Earth as Modified by Human Action*, first published in 1864. Marsh offered examples throughout history illustrating the indelible mark humans have left on the land:

> Man has too long forgotten that the Earth was given to him for usufruct alone, not for consumption, still less for profligate waste. The Earth is fast becoming an unfit home for its noblest inhabitant, and another era of equal human crime and human improvidence...would reduce it to such a condition of impoverished productiveness, of shattered surface, of climatic excess, as to threaten the depravation, barbarism, and perhaps even extinction of the species.

Marsh's book was widely read at the time in the United States and Europe and influenced those who later led the American conservation movement, particularly John Muir and Gifford Pinchot. In the post-Civil War period, however, national priorities were economic recovery and territorial expansion. Congress passed the Homestead Act in 1862 to encourage settlement of Western lands, and in the decades after the Civil War, many settlers moved west.

John Wesley Powell is remembered primarily as the one-armed adventurer who led the first expeditions by settlers through the Grand Canyon of the Colorado River. In his 1878 *Report on the Lands of the Arid Region of the United States, with a More Detailed Account of the Lands of Utah*, Powell countered the prevailing belief that dry western lands were suitable for settlement by small farmers, and that "the rain will follow the plow":

> The eastern portion of the United States is supplied with abundant rainfall for agricultural purposes... but westward the amount of aqueous precipitation diminishes in a general way until at last a region is reached where the climate is so arid that agriculture is not successful without irrigation.... To a great extent, the redemption of all these lands will require extensive and comprehensive plans, for the execution of which aggregated capital or cooperatives will be necessary. Here, individual farmers, being poor men, cannot undertake the task.

Powell's recommendations for federal policy, had they been followed, might have led to a different style of agriculture and land use in the West. However, Powell was unable to convince federal officials to adopt his recommendations. His report was ignored for three decades; it was not until 1902 that Congress passed the Reclamation Act and funded the dams and irrigation systems that Powell argued were essential if agriculture was to succeed in the arid areas of the West.

Conservation and Preservation: Theodore Roosevelt, Gifford Pinchot, and John Muir

By the end of the nineteenth century concerns about the depletion and misuse of the continent's resources led to the initiation of conservation activities by the federal government. The Forest Reserve Act of 1891 gave the President of the United States authority to set aside lands in the public domain so they could not be logged or otherwise developed, for the purposes of lessening the rapidity of depletion of the nation's forest base and protecting watersheds. During his presidency, Theodore Roosevelt set aside about 80 million acres of forested lands.

Roosevelt appointed Gifford Pinchot as head of the newly created Forest Service within the U.S. Department of Agriculture. Pinchot, who is credited with inventing the term "conservation," defined and implemented governmental policy in this area as "the use of the natural resources for the greatest good of the greatest number for the longest time." Pinchot, who studied forest management in Europe, argued for the scientific management of the nation's forest resources to assure a sustained yield for commercial use. In 1910 he cautioned against equating conservation with non-use: "The first principle of conservation is development, the use of the natural resources now existing on this continent for the benefit of the people who live here now."

In contrast to Pinchot's goals of scientifically managing natural resources for use, John Muir, a devoted hiker, naturalist, and founder of the Sierra Club, believed that some wilderness areas should remain untouched and unexploited for the benefit of all citizens. While he initially recognized the rights of both the progress of civilization and the protection of wilderness, he eventually took a clear stand as an advocate of nature untampered by man. His opposition to utilitarianism was clearly stated in a 1912 reaction to a plan, supported by Pinchot, to dam a particularly scenic valley as a reservoir for the city of San Francisco:

> That any one would try to destroy [Hetch Hetchy Valley] seems incredible, but sad experience shows that there are people good enough and bad enough for anything. The proponents of the dam scheme bring forth a lot of bad arguments to prove that the only righteous thing to do with the people's parks is to destroy them bit by bit as they are able... Dam Hetch Hetchy! As well dam for water tanks the people's cathedrals and churches, for no holier temple has ever been consecrated by the heart of man.

Though Muir's protest failed and the dam was built, public and political response to his arguments was instrumental in the establishment of the U.S. National Park Service, designed to preserve outstanding environmental features.

Aldo Leopold, often identified as the seminal figure in the development of modern environmental thought, was also a forester. He trained at the Yale School of Forestry and worked under Pinchot at the U.S. Forest Service. In his writings, Leopold spoke eloquently of the need for stewardship of the land and for a "land ethic":

> The first ethic [a reference to the Ten Commandments of the Judeo-Christian Bible] dealt with the relationship between individuals... Later accretions dealt with the relation between the individual and society... There is as yet no ethic dealing with man's relation to land and to the animals and plants which grow upon it... Land...is still property. The land-relation is still strictly economic, entailing privileges but not obligations. The extension of ethics to this third element in human environment is, if I read the evidence correctly, an evolutionary possibility and an ecological necessity...

Leopold's elegantly written *Sand County Almanac*, which was published a year after his death in 1947, has been one of the most influential books in contemporary thinking about man's relationship to nature.

The Emergence of a Popular Environmental Movement: Rachel Carson

World War II and the economic prosperity that followed in the 1950s pushed conservation issues off the national agenda. In the 1960s, however, federal clean air and clean water statutes, the Wilderness Preservation Act, and the National Wild and Scenic Rivers Acts were enacted, as attention grew to the necessity of preserving environmental quality. Charles Rubin has noted that the 1955 *New York Times Index* the word "environment" does not appear, although there were a few articles discussing local pollution issues. By the 1970s, the year of the first Earth Day, there were more than 85 articles indexed.[2]

There were a number of writers during the 1960s who expressed concern about the effects of growing urbanization and industrialization. Lewis Mumford, for example, in *Technics and Civilization* (1962), and microbiologist Rene Dubos in *So Human an Animal* (1968), warned that technological advances include human and ecological impacts that are not always beneficial. Rachel Carson's 1962 *Silent Spring*, however, is generally regarded as a pivotal event in launching the modern popular environmental movement. Carson, an ecologist, published a book for the popular market warning that pesticides were endangering bird populations. The book became a bestseller and was serialized in the *New Yorker*, reaching a wide audience. Other best-selling books, including Paul Ehrlich's 1968 *Population Bomb* and the Club of Rome's *Limits to Growth*, warned that population increases would place an ever-increasing strain on natural resources.

In comparatively rapid succession, Congress responded to popular concern about environmental quality by passing a number of major laws. In 1970, the National Environmental Policy Act was enacted and President Nixon established the Environmental Protection Agency, by consolidating a number of federal programs related to public health and safety. In 1970 the Occupational Safety and Health Administration was created; the Water Pollution Control Act, the Federal Insecticide, Fungicide, and Rodenticide Act, the Noise Control Act, the Coastal Zone Management Act, and the Marine Mammals Protection Act were all enacted in 1972. The Endangered Species Act was passed in 1973, the Safe Drinking Water Act in 1974, the Toxic Substances Control Act and the Resource Conservation and Recovery Act in 1976 – a flurry of environmental legislation at a pace that has not since been equaled.

Additional Resources

This is just a brief sketch of a rich and varied history of thought and action concerning the relationships between humans and nature in America. There are a number of books that provide more detail about the history of the conservation and environmental movements in America, some of which are listed below. In addition to the works cited above, authors such as James Fennimore Cooper, Herman Melville, John Steinbeck, and William Faulkner explored through their novels humanity's relationship to the natural world. Most of these works are available in public libraries, although it should be noted that many of the writings of Powell, Muir, Emerson, and Thoreau, among others, can be found in full text on the Internet. Following are some useful references:

de Crevecoeur, Hector St. John. *Letters from an American Farmer*. 1782. New York: Penguin American Library, 1981 (Reprint).
The letters were written in the colonies for publication in Europe, and offer insights into life and environment in early America.

Kalm, Peter (Pehr). *Travels in North America*. 1770. Dover Publications, 1987 (Reprint).
Kalm, a Swedish scientist, spent three years in North America studying plant life. This account of his travels records his observations of American agriculture, culture, and other aspects of American life.

Jefferson, Thomas. *Notes on the State of Virginia*. 1787. Reprinted widely.
In this classic work, Jefferson combined a detailed description of Virginia's rivers, land, and natural characteristics with an examination of its manners, culture, and institutions.

de Tocqueville, Alexis. *Democracy in America*. 1831. New York: Harper Collins, 1988 (Reprint).
Toqueville came to America in 1831 to study the effect of democracy on America. His expansive study examined the manners, political, family life, business, and agriculture of the country. This work, therefore, provides a detailed look of American life in the early 1800s.

Muir, John. *The Mountains of California*. New York: Penguin Books, 1985.
This reprint of a book by one of the founders of the American conservation movement includes reflections on and detailed descriptions of the wilderness of California's Sierra range.

Muir, John. *Our National Parks*. Boston: Houghton Mifflin, 1901. Madison: University of Wisconsin Press, 1981 (Reprint).
This is a collection of ten of Muir's essays, written to introduce Americans to the beauty and wonder of their national parks.

Library of Congress: The Evolution of the Conservation Movement 1850-1920
This online collection contains a documentary history of the conservation movement in America. The legislation creating Yosemite and Yellowstone parks can be found here, in addition to speeches and reports by all the major figures in the conservation movement, including Roosevelt, Pinchot, Muir, and Powell.
http://memory.loc.gov/ammem/amrvhtml/conshome.html

Pinchot, Gifford. *The Fight for Conservation*. New York: Doubleday, 1910.
In this collection of essays, Pinchot explains his belief that natural resources should be scientifically managed to prevent waste and preserve them for the common good. This book is available online, along with other classics works from conservationists, at *http://environment.about.com/newsissues/environment/msubeph.htm*

Pinchot, Gifford. *Breaking New Ground*. Washington, DC: Island Press, 1987.
Pinchot's autobiography, originally published in 1910, records the events of a period in which he was a key actor.

Powell, John Wesley. *Report on the Lands of the Arid Region of the United States, with a More Detailed Account of the Lands of Utah*. Washington, DC: Government Printing Office, 1879.
Major John W. Powell, geologist in charge of the Western region of the United States Geographical and Geological Survey, submitted this report to the Secretary of the Interior, with his recommendations on how the region could be made suitable for agriculture and grazing. The full text of this report can be found in the Library of Congress online collection, "The Evolution of the Conservation Movement 1850-1920" at
http://memory.loc.gov/ammem/amrvhtml/conshome.html

Powell, John Wesley. *Exploration of the Colorado River of the West and Its Tributaries*. Washington, DC: Government Printing Office. Reprinted as *The Exploration of the Colorado and Its Canyons*. New York: Penguin Books, 1987.
This is Powell's own account of his trips down the Colorado River. The full text of this report, and other writings by Powell, can be found at the University of North Texas' online archive at *http://www.library.unt.edu/gpo/powell/publications.htm*

Leopold, Aldo. *A Sand County Almanac*. New York: Ballatine Books, 1982.
In a series of nature essays, Leopold presents an elegant statement of the appropriate relationship between humans and their environments.

Flader, Susan L., and J. Baird Callicott, eds. *The River of the Mother of God and Other Essays by Aldo Leopold*. Madison, WI: University of Wisconsin Press, 1991.
Two eminent Leopold scholars present a selection of fifty of Leopold's essays written between his boyhood in 1904 and 1947, the year before his death.

DeVoto, Bernard, ed. *The Journals of Lewis and Clark*. Boston: Houghton Mifflin, 1953.
Meriwether Lewis and William Clark were sent by president Thomas Jefferson to explore the length of the Missouri River for its suitability for transporting people and goods across America. But he also instructed them to study the flora and fauna, the climate, and other natural characteristics of land. This book is recognized as the best version of the integrated journals of that expedition.

Botkin, Daniel. *Passage of Discovery: The American Rivers Guide to the Missouri River of Lewis and Clark*. New York: Berkley Publishing Group, 1999.
In a book that combines an historical and ecological study, Botkin retraces the route taken by Lewis and Clark to explore the Missouri River and uses their journals to see what the Western United States was like before it was changed by settlement and technology.

Cooper, James Fennimore. *The Pioneers*. New York: New American Library, 1964.
In several of his novels, Cooper explored the conflict between the growth of civilization in America and its frontier past. In *The Pioneers*, this conflict is typified by Natty Bumpo, who flees the frontier town for the woods, saying, "I'm formed for the wilderness; if ye love me, let me go where my soul craves to be ag'in."

Thoreau, Henry David. *Walden and Other Writings*. Edited by Joseph Wood Krutch. New York: Bantam Books, 1862.
This is Thoreau's account of his two years in a cabin in the woods (within walking distance of town) where he secluded himself to live simply, "be a neighbor to birds," and to see what the natural world had to teach him.

Thoreau, Henry David. *A Week on the Concord and Merrimack Rivers*. 1849. New York: Penguin Classics, 1998 (Reprint).
Thoreau describes here a leisurely sailing and rowing trip he took with his brother when they were both in their early twenties, in which he records his observations of life (human and wild) along the river.

Meadows, Donella et al. *The Limits to Growth*. New York: Universe Books, 1972.
In this study, a group of MIT researchers used a computer model to analyze trends in population growth, agriculture, industry, and other factors and predicted that environmental disasters would result. This study is often referred to as the Club of Rome report, because it was sponsored by the Club, an organization that was created by a wealthy Italian businessman to study global problems.

Dubos, René. *So Human an Animal*. New York: Charles Scribner's Sons, 1968.
Dubos, a microbiologist and one of the leading researchers in the development of antibiotic drugs, argued that science and technology can serve fundamental needs, but that humans must exercise self-restraint to avoid damage to ecological systems.

Dubos, René. *A God Within*. New York: Charles Scribner's Sons, 1972.
In this work Dubos tells how we are shaped by our environment and how we can control forces such as urbanization and technology to meet human needs.

Ehrlich, Paul. *The Population Bomb*. Rivercity, MA: Rivercity Press, 1975.
In this 1968 book, biologist Ehrlich predicted that rapid population growth would lead to worldwide famine and environmental degradation.

Rachel Carson. *Silent Spring*. Boston: Houghton Mifflin, 1962.
This book played a pivotal role in the environmental movement of the 1960s. The book's title refers to the book's first chapter in which Carson describes a world so devastated by synthetic pesticides that birds are silenced.

Nash, Roderick. *Wilderness and the American Mind.* Third Edition. New Haven: Yale University Press, 1982.
Nash provides a thorough history of how Americans have regarded the natural world and their relationship to it.

Worster, Donald. *Nature's Economy: A History of Ecological Ideas.* New York: Cambridge University Press, 1985.
Ecology is a relatively new science with a fascinating history. This is an interesting, but challenging, account of the development of ecology.

Miller, Perry. *Errand Unto the Wilderness.* Cambridge, MA: Harvard University Press, 1956.
Miller explores the inner tensions in how Americans have historically regarded the natural world, from early Puritans' views of America as "a hideous and desolate wilderness full of wilde beastes and men," to the romantic view of artists such as Thomas Cole and writers such as Emerson and Thoreau.

Opie, John. *Nature's Nation: An Environmental History of the United States.* Philadelphia: Harcourt Brace Publishers, 1998.
This is an examination of the course of American history from an environmental point of view that makes connections between historical events and environmental conditions.

Matthiessen, Peter. *Wildlife in America.* New York: Viking Press, 1959.
This volume chronicles American wildlife from the time of European arrival to the mid-1900s.

Nash, Roderick, ed. *American Environmentalism: Readings in Conservation History.* Third Edition. New York: McGraw-Hill, 1990.
This is a collection of the most prominent readings from the conservation and environmental movements over the past 160 years.

Hays, Samuel P. *Beauty, Health, and Permanence: Environmental Politics in the United States*, 1955-1985. New York: Cambridge University Press, 1987.
Hays, a noted conservation historian, focuses his attention on the environmental movement of the recent past, with consideration of what triggered it, what happened, and what it means in the long term.

Hays, Samuel P. *Conservation and the Gospel of Efficiency: The Progressive Conservation Movement, 1890-1920.* Cambridge, MA: Harvard University Press, 1959.
This book is a scholarly history of the early conservation movement.

Cronon, William. *Changes in the Land: Indians, Colonists, and the Ecology of New England.* New York: Hill and Wang, 1983.
Cronon examines the record of the relationship between humans and nature in colonial New England.

Sears, Paul B. *Deserts on the March.* Norman, OK: University of Oklahoma Press. 1935. Reprinted 1988 by Island Press.
An historical account of desertification, historically and globally. This book was widely read in the 1930s.

Worster, Donald. *Dust Bowl: The Southern Plains in the 1930s.* New York: Oxford University Press, 1979.
A scholarly history of causes and effects of the Dust Bowl in the Southern Great Plains of the United States.

Turner, Frederick Jackson. *The Frontier in American History.* New York: Henry Holt and Co., 1920. Reprinted 1986 by University of Arizona Press.
A collection of Turner's essays on the American frontier.

Stegner, Wallace. *Beyond the Hundredth Meridian: John Wesley Powell and the Second Opening of the West.* New York: Penguin Books, 1992.
This book, a biography of John Wesley Powell, is also an excellent history of America in the nineteenth century and the settling of the West.

Caldwell, Lynton Keith. *The National Environmental Policy Act: An Agenda for the Future.* Bloomington, IN: Indiana University Press, 1998.
Caldwell reviews the evolution of the National Environmental Policy Act from its passage in 1969 to the present day. Caldwell, who was the principle author of the Act, argues that its potential as a catalyst for environmental improvement has so far been largely unrealized because of lack of political commitment.

Landy, Marc K., Marc J. Roberts, and Stephen R. Thomas. *The Environmental Protection Agency: Asking the Wrong Questions.* New York: Oxford University Press, 1990.
The authors trace the history of the U. S. Environmental Protection Agency from its creation in 1970 through the late 1980s.

Background

The Earth is almost certainly the only planet in the solar system capable of supporting life, because it is uniquely positioned to hold a vast quantity of water in liquid form. Water is essential for life because it is the ideal solvent for the chemical reactions that occur in living cells. Water also regulates the temperature of the planet at a tolerable level and cycles essential nutrients from land, air, and living things. Although 71 percent of the Earth's surface is covered with water, less than 1 percent of the total is fresh water available for use. Moreover, fresh water is not evenly distributed across land surfaces and a number of heavily populated countries are in arid lands where fresh water is scarce.

> Going up that river was like traveling back to the earliest beginnings of the world, when vegetation rioted on the earth and the big trees were kings. An empty stream, a great silence, an impenetrable forest.
>
> ★Joseph Conrad, **Heart of Darkness**★

Philosophies guiding management of water supplies changed during the 20th century. Until the second half of the 1900s, water management was governed by the goal of moving water to where it was most needed, particularly for irrigation of agricultural lands. Rivers, lakes, and other bodies of water were also used to carry away wastes from municipal and industrial uses, because it was believed that the pollutants would disperse in the water. In the past three decades, the focus of water management has shifted to considerations of municipal, agricultural, and industrial supplies, water quality, and protection of aquatic ecosystems.

What are the issues?

All communities need an adequate supply of fresh water for drinking, agriculture, and industrial uses, with its quality being as important as its quantity. Water-borne diseases remain a significant source of mortality in many developing countries. The United States has made significant progress over the last thirty years in improving water quality in its lakes, rivers, and streams and in reducing direct discharges of pollutants. However, indirect sources of pollutants, including both nutrients and toxic substances that are carried by flowing water into larger bodies of water, are more difficult to control.

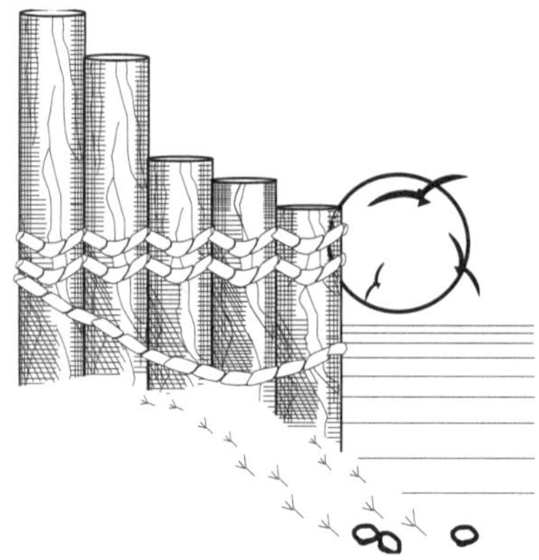

Rivers, lakes, oceans, and bays also support rich ecosystems. Changes in the environment due to human action can have a detrimental impact on these ecosystems. When dams, for example, divert water for agricultural irrigation, they affect land below and above the dam. In the past few decades, research has demonstrated the importance of coastal ecosystems such as wetlands, mangroves, and coral reefs and their susceptibility to damage from human activities on the coast. Fishing provides a valuable source of nutrients for many populations, but depletion of fish populations is a problem in some areas. Efforts to prevent overfishing, ocean dumping, and other destructive actions are the subject of international treaties, but these treaties can be difficult to enforce and violations difficult to detect because of the vast size of the oceans.

What are the uncertainties?

Far more of the Earth's surface is covered by ocean than by land, but much of it remains an unexplored frontier. New technologies developed only in the last few decades, such as remote sensing and deep-sea exploration vehicles, have enhanced scientists' ability to study the interactions between ocean and atmosphere, but questions remain. For example, neither how the oceans help to cycle nutrients nor their role in regulating global temperatures is completely understood (see Global Climate Change).

Sources of fresh water, including lakes, rivers, and streams, are more accessible for research, so more is understood about these ecosystems. The relationship of the water quality of a river or lake to changes in land use in the watershed that feeds it is well known, but regulating or managing the diverse mix of human activities that occur over such broad areas, and often over many governmental jurisdictions, has proven to be difficult. Agriculture, lawn care, snow removal practices, and actions of millions of individuals that affect water quality are difficult to address.

What are the risks and tradeoffs?

Water is essential to life and conflict between different goals often occurs. For example, efforts to ensure adequate water supply by damming or modifying the flow of rivers can affect water quality and ecosystems. Inadequate sanitation and poor water quality are two of the most serious health problems facing developing countries. A rapidly urbanizing population increases potential conflicts between agriculture and urban uses for the water supply. Water demand from agriculture needed to feed a growing population can deplete ground and surface water reservoirs, with potential impacts on water quality and ecosystem health.

Basic Concepts

The flow of water through the atmosphere, biosphere, lithosphere, and hydrosphere is called the hydrologic, or water cycle. This cycle is explained in most general science textbooks; Earth science textbooks may provide greater detail. The study of water in its various reservoirs (oceans, rivers, streams, aquifers), in addition to the ecosystems supported by these reservoirs, is the work of a number of scientific disciplines, from hydrology to marine biology, and textbooks in these specialized fields may be useful for further research. Following are some of the important concerns related to water supply and quality and some basic resources for research on water issues.

Moore, John et al. *Groundwater: A Primer.* Alexandria, VA: American Geological Institute, 1995.
Designed for teachers and students, this booklet describes fundamental concepts of groundwater occurrence, development, quality, use, and protection. The booklet can be ordered online at *http://www.agiweb.org*

Leopold, Luna B. *Water, Rivers, and Creeks.* Sausalito, CA: University Science Books, 1997.
This is an easy to read, thorough primer on hydrology by the former Chief Hydrologist of the U.S. Geological Survey. Leopold, son of conservationist Aldo Leopold, explains in simple language how water moves through water, air, and soil.

Pielou, E.C. *Fresh Water.* Chicago: University of Chicago Press, 1998.
This is a natural history of water written by an ecologist. In clear, simple language, the author explains basic hydrological concepts such as the water cycle, how streams originate and grow, how lakes are formed, and how to measure the velocity of water.

Leopold, Luna B. *A View of the River.* Cambridge, MA: Harvard University Press, 1994.
Leopold focuses on rivers in this book, discussing how rivers are formed, make channels, flood, and fill with silt. The book is fairly advanced, but written for a lay audience.

National Council for Science and the Environment: National Library for the Environment
Congressional Research Service reports and other reference materials are available on water topics including desalination, land use, pollution control, and water supply to urban and rural areas. *http://www.cnie.org*

Water Supply

In the United States, fresh water supplies come from both surface water reservoirs and groundwater. Almost 1 percent of water (including saltwater) on Earth is stored as groundwater, which is water contained in spaces between particles—in pores, cracks and fissures—in bedrock and soils. By volume, the Earth has 40 times as much groundwater as surface fresh water, so it is an important source of water. Groundwater from wells has been used as a source of water for domestic uses and irrigation for thousands of years. The development of electric pumps, however, has made it possible for water to be drawn at a faster rate than it is replenished through natural processes (rain and snow), raising questions about the adequacy of the supply in some areas.

Golley, Frank B. *A Primer for Environmental Literacy.* New Haven: Yale University Press, 1998.
Golley includes a useful chapter on watersheds in this basic primer on the Earth's systems. He explains the dynamics of water in the watershed (with diagram) and provides a brief, clear introduction to the biochemical flows through the watershed.

U.S. Geological Survey: Water Resources of the United States
This online publication lists water use, water quality and flow data, and includes a groundwater atlas of every major aquifer in the United States. Available at *http://h2o.usgs.gov* A copy of *The Ground Water Atlas of the U.S.* can be ordered at 1-888-Ask-USGS.

"Water: The Power, Promise, and Turmoil of North America's Fresh Water." ***National Geographic Special Edition*, November 1993.**
This special issue focuses on water supply and water quality, including articles on topics such as the Mississippi River, the Age of Dams, and a proposal to grow rice in California.

U.S. Geological Survey: High Plains Aquifer Resources and Information
One important source of water for agriculture in the United States, the Ogallala Aquifer, has been a cause of concern because it has been significantly depleted in some areas, particularly Texas. The Ogallala or High Plains aquifer lies under states from South Dakota to Texas and is the primary source of water in this region for irrigation.
http://ne2odnelnc.cr.usgs.gov/highplains/glm.hpactivities.html
A case study on the Ogallala can be found at
http://agecon.lib.umn.edu/aaea99/sp99te01.pdf

Frederick, Kenneth. "America's Water Supply: Status and Prospects for the Future." ***Consequences*, Spring, 1995.**
The author presents a clear, detailed overview of issues related to water supply in the United States. In particular, he explains how all the users (industrial, municipal, and agricultural) of water in a watershed become increasingly interdependent, and how each type of water use can affect the quantity and quality of water available to others. This article is available online at
http://www.gcrio.org/CONSEQUENCES/spring95/Water.html

Fredrick, Kenneth et al. "Economic Values of Fresh Water in the United States: A Report to the Electric Power Research Institute." Resources for the Future, December 10, 1996.
This is a comprehensive report on the value of fresh water in the United States. Available at *http://www.rff.org/reports/summaries/watervalues.htm*

WATER SUPPLY IN THE WESTERN UNITED STATES

Many areas in the Western United States are arid, and, since settlement began, finding and transporting an adequate water supply to support agriculture and municipal uses has been a constant challenge. Most of the major rivers in the West have been dammed to provide water for irrigation or to supply water to cities such as San Francisco and Los Angeles. This has had significant impacts on water quality in the surrounding ecosystems, including many fish breeding grounds.

Western Water Policy Review Advisory Commission
Congress established this commission to evaluate federal water policy in the West. The commission includes on its site a number of research reports, most notably, "Water in the West: The Challenge for the Next Century." This lengthy report covers water use, hydroelectric power, and other natural resources in the Western United States.
http://www.den.doi.gov/wwprac/

Solley, Wayne. "Estimates of Water Use in the Western United States in 1990 and Water-Use Trends 1960-1990." U.S. Geological Survey, August 1997.
The Western Water Policy Review Advisory Commission found agriculture to be the single greatest category of water use in 1990, accounting for almost 78 percent of all water withdrawals, although groundwater withdrawals have declined since 1975 due to more

efficient irrigation systems and the introduction of crops that use less water. This report may be obtained by calling 1-888-ASK-USGS.

Department of the Interior: Bureau of Reclamation
This federal agency was established to develop irrigation projects for the Western United States. Their site includes a history of the agency and information about water supply in western rivers and reservoirs.
http://www.usbr.gov/main/index.html

California Coastal Commission. "Seawater Desalinization in California." October 1993.
This report explains how desalination plants remove salt and other substances from seawater. Available at *http://www.ceres.ca.gov/coastalcomm/pubs.html*

Joseph, Patrick. "The Battle of the Dams." *Smithsonian*, November 1998, pp. 48-61.
The author examines the controversies surrounding dam projects. Although dams provide clean power, irrigation, and flood control, they also disrupt local ecosystems.

DeLong, James. "Dam Fools." *Reason* 29 (April 1998): 40-47.
DeLong criticizes federal spending on dam construction for economic and environmental reasons. He reviews data indicating that dams provide taxpayer-subsidized water to a few major users. *http://www.reason.com/9804/fe.delong.html*

Foundation for Water and Energy Education: Environment
This page lists information on the Colorado River Watershed and explains how hydroelectric projects affect their surrounding areas. *http://www.fwee.org/environment.html*

Anderson, Terry, and Pamela Snyder. *Water Markets: Priming the Invisible Pump.* Washington, DC: Cato Institute, 1997.
These authors argue that water scarcity in the West is exacerbated by the fact that water supplies are subsidized by state and federal governments and users of the water have few incentives to conserve. They believe that water markets would have economic and environmental benefits. This book can be ordered from *http://www.cato.org*

Worster, Donald. *Rivers of Empire: Water, Aridity, and the Waters of the American West.* New York: Oxford University Press, 1992. (Reprint)
This is a history of the emergence and effects of large irrigation projects in the West.

Powell, John Wesley. *Report on the Lands of the Arid Region of the United States, with a More Detailed Account of the Lands of Utah.* Washington, DC: Government Printing Office, 1878.
After an extended survey of Western lands, John Wesley Powell warned administration officials that the lack of rainfall in the desert areas of the West would doom the small farms that were envisioned by the 1862 Homestead Act and that irrigation would be necessary. The full text of this report can be found on the Library of Congress' online collection, "The Evolution of the Conservation Movement 1850-1920," at
http://memory.loc.gov/ammem/amrvhtml/conshome.html

Reisner, Marc. *Cadillac Desert: The American West and its Disappearing Water.* New York: Penguin Books, 1993.
A provocative, opinionated, and interesting history of how California, the Bureau of Reclamation, and the Corps of Engineers remade the West by damming its rivers. The author explores the early history of Western settlement and the mistaken belief of the

time that "rain follows the plough." The author's account of the failure of several dams is absorbing.

WATER QUALITY

As critical as an adequate supply of water is its quality. Water-borne diseases have historically been a significant cause of mortality and continue to be in many undeveloped countries. Water quality in the U.S. has improved significantly in the last thirty years as local, state, and federal legislation have prohibited direct discharges of wastes into rivers, lakes, and streams (called point pollution). The issue of indirect sources of pollutants, or non-point pollution, remains problematic. As water from rain and melted snow flow into a watershed, the water carries with it nutrients and wastes. These nutrients increase growth of algae in bodies of water, and as the increased population of algae decomposes, oxygen is taken up, in a process termed eutrophication. This process, which transforms bodies of water from oxygen-rich to oxygen-deficient, degrades the habitat for fish. Because nutrients are transported through the watershed from a number of diverse sources, municipal and agricultural, it has proven more difficult to identify and manage these sources.

Laws, Edward. *Aquatic Pollution: An Introductory Text*. Second Edition. New York: John Wiley & Sons, Inc., 1993.
This is an upper-level text that provides comprehensive and systematic coverage of pollution in marine and fresh water systems. Four chapters are devoted to ecological principles and basic toxicology.

Environmental Protection Agency, Office of Water
This site provides a map that permits the user to locate a watershed and retrieve data regarding its water quality. *http://www.epa.gov/OW/*

North Carolina State University, Water Quality Group: WATERSHEDS
This page has extensive information about water quality, point, and non-point source pollution, including definitions, case studies, and detailed explanations about measuring water quality. The site is intended for field professionals, but would also be useful for upper-level students.
http://h2osparc.wq.ncsu.edu/

Knopman, Debra, and Richard Smith. "Has U.S. Water Quality Improved? Twenty Years of the Clean Water Act." *Environment*, January/February 1993, pp. 16-20, 34-41.
The authors argue that twenty years after its passage, policy makers know little about the effectiveness of the Clean Water Act. Local studies have been useful for water quality management, but these results are difficult to aggregate nationally.

National Acid Precipitation Assessment Program. *1990 Integrated Assessment Report*. Washington, DC: Government Printing Office, 1991.
This report was the result of a ten-year study on the causes and effects of acid precipitation. The study found that the effects of acidic precipitation were more modest than originally hypothesized and that, although acid rain had caused the acidification of some lakes in the northeastern United States and Canada, the negative effects were limited to a relatively few lakes that lack adequate buffering capacity (See also Air Quality).

Baker, Lawrence et al. "Acidic Lakes and Streams in the United States: The Role of Acidic Deposition." ***Science*** **252 (May 24, 1991): 1151-1154.**
This article reviews the results of a major scientific study conducted to determine the degree to which acidic conditions in U.S. lakes and streams are caused by acidic deposition.

Living Lakes
This conservation and education organization has information on a number of lakes around the world, including Mono Lake in California. The site describes the characteristics of individual watersheds, examining the biological life, human history around the lakes, and conservation efforts. *http://www.livinglakes.org/*

Great Lakes Information Network: Environment of the Great Lakes Region
This page links to information on water pollution, water quality, watershed management, and conservation efforts specific to the area. It also provides links to other related public and private organizations outside the Great Lakes Region that examine water quality issues. *http://www.great-lakes.net/envt/envt.html*

Makarewicz, Joseph, and Paul Bertram. "Evidence for the Restoration of the Lake Erie Ecosystem." ***BioScience*** **41 (April 1991): 216-223.**
Lake Erie, one of the most noted cases of water pollution in the United States, has seen significant improvements in water quality. This article argues that the large reduction in the amount of industrial dumping of phosphorus accounts for the decrease in harmful phytoplankton and the subsequent rebound of oxygen levels and fish populations.

Raloff, Janet. "Drugged Waters." ***Science News*** **153 (March 21, 1998): 187-189.**
Water quality can be affected by many individual activities that are difficult to regulate. This article discusses various proposals for identifying and controlling the amount of pharmaceuticals found in water supplies, not from direct industrial pollution, but rather from human wastes.

Malakoff, David. "Death by Suffocation in the Gulf of Mexico." ***Science*** **281 (July 10, 1998): 190-192.**
This article addresses the theory that nitrogen, accumulated in the Mississippi Basin from nitrogen-rich fertilizer carried off from farmland into the Mississippi River by melting snow and spring rains, is the cause of hypoxia, or oxygen-poor water pockets, which reduce populations of fish and shrimp which are critical to Louisiana's economy.

Boyle, Lance. "The Life or Death of the Salton Sea?" ***Smithsonian*****, June 1996, pp. 87-97.**
The author describes plans to restore the Salton Sea in Southern California, an inland lake created when the Colorado River broke out of its channel. The sea is now threatened by increasing salinity and agricultural runoff.
See *http://www.smithsonianmag.si.edu/smithsonian/issues96/jun96/salton.html* for additional resources on this topic.

Parlange, Mary. "Parasites in the Pipes." ***BioScience*** **49 (May 1999): 360-364.**
Water-borne parasites, *Cryptosporidium*, are believed to be ubiquitous in the U.S. water supply. No reliable method exists to detect the pathogen in water samples and it is difficult to treat. Although the health effects of *cryptosporidiosis* are relatively minor for

most, infections can be fatal to individuals with weakened immune systems such as those suffering from HIV and AIDS.

Guynup, Sharon. "The Cell from Hell." ***Popular Science*****, June 1999, pp. 66-70.**
Guynup discusses the toxic microbe known as *Pfiesteria piscicida* which has been blamed for fish kills along the east coast of the United States. Depending upon its prey, *Pfiesteria* can morph into twenty-four different body forms. Although little is known about the chemical composition of *Pfiesteria*, a twelve-minute field test has been developed to detect the substance in water.

History of Municipal Water Systems

Narragansett Bay Commission: History
This page traces the development of the water treatment facility in Providence, Rhode Island, from the 1854 cholera epidemic that motivated the city to treat drinking water, to the present state of the facility. *http://www.narrabay.com/org/history.html*

Ausbel, Jesse H., and Robert Herman. ***Cities and Their Vital Systems: Infrastructure Past, Present, and Future*****. Washington, DC: National Academy Press, 1988.**
Philadelphia responded to a yellow fever epidemic in the late 1700s by building the first municipal water system, these authors report; the first modern urban sewage system was built in Brooklyn in 1857. This is an expensive publication, but it can be viewed at *http://www.nap.edu* (search by title).

Mary Gayman. "History of London Sewers." ***Cleaner Magazine*** **(Reprinted by SwopNet Engineering Databank).**
The term "sewer" meant "seaward" in Old English, and ditches were dug along city streets to drain human wastes into the Thames and ultimately to the sea. This site provides a fascinating history of London's sewers, including a proclamation by King Henry VIII making every household responsible for cleaning the sewers in front of their house, and Sir Thomas Crapper's invention of the flush toilet.
http://www.swopnet.com/engr/londonsewers/londontext1.html An equally fascinating "History of Paris Sewers" is available at
http://www.op.net/~uarts/krupa/alltextparis.html

Water in Developing Countries

In many parts of the world, lack of an adequate supply of drinking water is a serious environmental and health issue. It is estimated that 5 to 10 million deaths a year, mostly of small children and elderly, from cholera and other diseases are caused by poor water quality and sanitation.[3] Rapid urbanization in many developing countries has outpaced their capacities to build an infrastructure to meet basic human needs for clean water and sanitation. Water-borne illnesses remain a significant cause of mortality for many countries.

World Health Organization: Water Sanitation and Health
This page offers information on water quality and supply in developing countries and on the negative health effects of poor water quality.
http://www.who.int/water_sanitation_health/index.html

Glass, R., M. Libel, and A. Brandling-Bennett "Epidemic Cholera in the Americas." ***Science*** **265 (June 12, 1992): 1524-1525.**
A widespread cholera epidemic in the early 1990s underlines the fact that water quality problems still exist in much of the developing world. This article discusses the ways that cholera is transmitted, and explains some new preventative measures, treatments, and vaccinations for this often-fatal disease.

Okun, Daniel. "A Water and Sanitation Strategy for the Developing World." ***Environment,*** **October 1991, pp. 16-20, 38-43.**
The author explains why it is not a good idea to drink the water in many countries. Rapid urban growth has exacerbated the problem of providing a safe water supply for many countries. The author argues that lack of funding is less of a problem than building the necessary infrastructure and educating engineers and managers to run the systems.

Nakamura, Masahisa. "Preserving the Health of the World's Lakes." ***Environment,*** **June 1997, pp. 16-20, 34-40.**
Nakamura discusses damage done to the world's lakes by agricultural runoff, acidification, inadequate wastewater treatment, and the absence of oxygen infusion. Highlighting discussions from a 1995 conference held near Lake Kasumigaura in Japan, Nakamura believes the trend of lake degradation can be reversed.

Baskin, Yvonne. "Losing a Lake." ***Discover,*** **March 1994, pp. 72-81.**
Africa's Lake Victoria, the second largest lake in the world, is seriously choked by algae, and half of its native fish have been driven to extinction by the Nile perch, which, although it has disrupted the ecosystem, is an important source of protein for surrounding populations.

Gleick, Peter et al. "Water, War & Peace in the Middle East." ***Environment,*** **April 1994, pp. 6-15, 32-42.**
The authors discuss the competing social, political, and geographic variables that have complicated obtaining agreement about scarce water supplies in the Middle East. The authors believe that market-based water banks might foster greater efficiency and ease regional tensions.

Vesilind, Peter. "Water: The Middle East's Critical Resource." ***National Geographic,*** **May 1993, pp. 38-70.**
Political tensions and inefficient practices have hindered management of the Middle East's scarce freshwater supplies. This lengthy article discusses obstacles faced by countries in this arid region where water is becoming almost as precious a resource as oil.

Ocean And Coastal Waters

Most of the world population lives within 100 kilometers of the sea, but there is much about the ocean that scientists are just beginning to understand. Satellites, undersea exploration vehicles, and other technological developments have revealed new information and raised more questions. It is clear that the great ocean currents affect the Earth's temperature. Oceans participate in creating weather, redistribute heat energy across the Earth surface, and cycle nutrients such as carbon dioxide through Earth's systems. The oceans are also a vast global commons. International treaties govern activities, such as dumping of wastes, overexploitation of fisheries, and hunting of marine creatures such as whales, but these treaties have proved to be difficult to enforce.

Skinner, Brian J., and Stephen C. Porter. *The Dynamic Earth: An Introduction to Physical Geology.* New York: John Wiley & Sons, Inc., 1992.
This basic textbook describes the ocean forces that shape the coast, how beaches are formed and transformed, and how barrier islands are formed.

National Oceanic and Atmospheric Administration: National Ocean Service
This site provides information on coastline health, including a *State of the Coast Report.*
http://www.nos.noaa.gov/

Virginia Institute of Marine Science: The Bridge – Ocean Sciences Education Teacher Resource Center
This is a marine science education site that includes information on marine organisms, habitats, use, and management.
http://www.vims.edu/bridge/

Scripps Institution of Oceanography Library
Scripps provides an extensive list of links to oceanographic resources on the Internet. Also included on their site is a list of oceanographic mapping sources providing images of wave heights, ocean currents, El Niño patterns, surface temperature, salinity, and other oceanographic data. *http://scilib.ucsd.edu/sio/*

World Resources Institute: Coastal and Marine Resources
These pages cover a wide range of marine issues including biodiversity, threats to marine resources, and coastal management. *http://www.wri.org/wri/biodiv/marihome.html*

"The Oceans." *Scientific American Presents*, Fall 1998.
This special edition of *Scientific American* includes articles on El Niño currents, marine habitats, and coral reefs, ocean law, and fisheries, among other topics. Interesting bits of information are also here, including the origin of water on Earth and how the oceans got their names.

"Oceans and Coastal Resources: A Briefing Book." Congressional Service Report for Congress, No. 97-588 ENR, May 30, 1997.
This extensive compilation includes discussions of marine fisheries, coastal development, current policy issues, information on the coastal environment, and ocean and coastal data. Available at *http://www.cnie.org/nle/mar-20/*

Australian Institute of Marine Science: Project Net
Designed to aid students undertaking marine science projects, topics on this site range from the Great Barrier Reef and dangerous marine animals to tsunami waves and mangroves. Most topics link to additional information and list books for further reading.
http://www.aims.gov.au/pages/research/project-net/apnet.html

Pennisi, Elizabeth. "Sorcerers of the Sea: Making Microbes Do Our Dirty Work." *BioScience* 46 (April 1996): 236-239.
Bioremediation, the use of microbes to break down or remove pollutants from the environment, has been the focus of researchers who are looking for inexpensive and efficient ways to remove pollution from the world's oceans. By stimulating the growth of "scrub-

bing" bacteria already in the ocean, or by placing genetically altered microbes in a polluted area, it is possible to degrade toxins several times faster than normal.

Lawrence, David. ***Earth Science: A Graphing Calculator Approach.*** **Fairfield, NJ: Pencil Point Press, Inc., 1999.**
This manual demonstrates using a graphing calculator for environmental science-related exercises, including simulating a map of the ocean floor.

Pilkey, Orrin, and Katherine L. Dixon. ***The Corps and the Shore.*** **Washington, DC: Island Press, 1996.**
An opinionated, interesting history of the U.S. Corps of Engineers, which began modifying the shoreline of the Eastern U.S. during the Revolutionary War. The authors report how the Corps, responding to public pressures to save beaches, has fought a losing battle with the natural forces of wind and water that shape the land.

MARINE ECOSYSTEMS

Agricultural runoff, fishing, and other human activities can affect coastal ecosystems including coral reefs, mangroves, and estuaries such as the Chesapeake Bay. Major conservation programs are underway in many areas, including the Everglades. Protection of coral reefs remains a serious concern. These rich ecosystems are usually in international waters, and although monitoring has increased, illegal fishing and the use of cyanide or explosives to stun fish are severely degrading reef systems in the South China sea and other areas.

Botsford, Louis W., Juan Carlos Castilla, and Charles H. Peterson. "The Management of Fisheries and Marine Ecosystems." ***Science*** **277 (July 25, 1997): 509-515.**
Fish account for about 19 percent of total human consumption of animal protein, but a number of important fisheries are fully exploited or overexploited. The authors argue that fisheries managers in many countries often succumb to political pressures to increase catch limits, even when physical influences such as *El Niño* have affected marine ecosystems.

Chadwick, Douglas. "Coral in Peril." ***National Geographic,*** **January 1999, pp. 30-37.**
Chadwick reports on the health of coral reef systems and natural and human factors that have affected them. The article is accompanied by amazing photographs of this ecosystem and surrounding sea life.

National Park Service: Everglades Ecosystem
This U.S. agency site provides information on the Everglades' habitats, inhabitants, geology, and history. *http://www.nps.gov/ever/eco/*

Asian Institute of Technology, Thailand: Mangrove Conservation and Development
This site explains the mangrove ecosystem, how it has been used, the current state of mangrove stands worldwide, and conservation measures. The site includes a bibliography and link to other sites. *http://www.agri-aqua.ait.ac.th/Mangroves/main.html*

National Oceanic and Atmospheric Administration: Sanctuaries and Reserve Division
This site provides links to the National Estuarine Research Reserve Sites, including sites such as the Florida Keys, Monterey Bay, and the Channel Islands. Each of these sites

include information about the flora and fauna of the reserve and research conducted on site. *http://www.sanctuaries.nos.noaa.gov/*

WETLANDS

Swamps and other wetland areas were once regarded as wasted land and millions of federal dollars were spent to drain and fill them. The Swamplands Acts in the mid-1800s, for example, transferred millions of acres of land to states to be drained for cropland. Now it is understood that wetlands, which are areas in which water saturates or floods the soil much of the time, have important ecosystem functions. These areas provide a fertile habitat for waterfowl, plants, and wildlife and they also filter pollutants, reduce flooding, and buffer coasts and waterways.

Pielou, E.C. *Fresh Water*. Chicago: University of Chicago Press, 1998.
This book, which is a natural history of water written by an ecologist, provides a clear explanation of what a wetland is and the differences among various types of wetland ecosystems. For example, we learn that a bog is a type of wetland that is supplied by rainwater; its water contains few nutrients to support fish and plant life. A fen, by contrast, is fed by groundwater seepage; its water picks up nutrients as it has moved through the soil, supporting a greater variety of plants and fish.

Krajick, Kevin. "The Riddle of the Carolina Bays." *Smithsonian*, September 1997, pp. 45-56.
Along the eastern coastal plain of the United States lie a series of "bays," or oval-shaped wetland depressions. The regularity and shape of these formations were discovered only after satellite images of the East Coast were available. Krajick describes the diversity and uniqueness of bay wildlife habitats.

U.S. Fish and Wildlife Service: National Wetlands Inventory
This site provides access to wetland maps, data, and reports, including information about plants and animals that live in wetland habitats. *http://www.nwi.fws.gov/overiew.htm*

Vileisis, Ann. *Discovering the Unknown Landscape: A History of America's Wetlands*. Washington, DC: Island Press, 1999.
This book provides a detailed historical look at how Americans have regarded wetlands, from the early settlers' fear of these "dismal" areas to current policies aimed at restoration and protection.

National Research Council. *Wetlands: Characteristics and Boundaries*. National Academy Press, 1995.
This publication reviews how wetlands are classified, their ecological importance, and current policies. Available to read or purchase at *http://www.nap.edu*

Ewel, Katherine. "Multiple Demands on Wetlands." *BioScience* 40 (October 1990): 660-666.
Cypress swamps in Florida are a source of lumber, which has increased the value of these wetlands, but cypress lumber harvests have removed nesting cavities and cover for birds and mammals. Ewel examines how competing uses for wetlands can be managed.

Kusler, Jon, William Mitsch, and Joseph Larson. "Wetlands." ***Scientific American*****, January 1994, pp. 64-70.**
The authors stress the cyclical relationship between wetland ecosystems and their surrounding environment and suggest that management strategies should simulate natural processes as closely as possible.

Gillette, Becky. "Wetlands Treatment: Getting Close to Nature." ***BioCycle*** **37 (October 1996): 74-76, 78.**
Gillette contends that constructed wetlands should be designed to mimic those existing in nature. Instead of one large system, constructed wetlands should be a series of cells and channels accommodating a diversity of wildlife.

Roberts, Leslie. "Wetlands Trading Is a Loser's Game, Say Ecologists." ***Science*** **37 (June 25, 1993): 1890-1892.**
The author contends that projects by developers to replace natural wetlands with artificial ones have been unsuccessful. According to the author, these mitigation projects are driven more by economic and political interests than by scientific research.

Zinn, Jeffrey, and Claudia Copeland. "Wetlands Issues." Congressional Research Service Report for Congress, No. 97014, August 3, 1999.
This brief discusses the importance of wetlands historically and how policy has changed as knowledge about wetland significance increased. Available at
http://www.cnie.org/nle/wet-5.html

Kusler, Jon. "Wetlands Delineation: An Issue of Science or Politics?" ***Environment*****, March 1992, pp. 6-11, 29-37.**
This article examines criticisms of 1991 U.S. wetlands protection legislation and traces the history of wetlands protection politics from the early 1970s to the early 1990s.

LEGISLATION AND POLICY

The Clean Water Act (33 U.S.C. sec. 121)
This is the major law regulating water quality in the United States. The Clean Water Act established a framework for regulating discharges of water pollutants in the United States. *http://www4.law.cornell.edu/uscode/33/ch26.html*

The Safe Drinking Water Act (42 USC sec. 300f)
This legislation establishes purity standards for drinking water in the United States.
http://www4.law.cornell.edu/uscode/42/300f.html

The Oil Pollution Act (33 USC sec. 2702 to 2761)
This act established a fund financed by a tax on oil to clean up spills when the responsible party lacks the necessary means and requires that spill contingency plans be submitted to the Environmental Protection Agency. *http://www.epa.gov/region5/defs/html/opa.htm*

Copeland, Claudia. "Water Quality: Implementing the Clean Water Act." Congressional Research Service Report for Congress, No. 89102, August 6, 1999.
This brief for Congress summarizes the Clean Water Act and its amendments, discusses recent policy developments, and reviews implementation procedures. Available at
http://www.cnie.org/nle/h2o-15.html

United Nations: Ocean and Law of the Sea Home Page
The United Nations Convention on the Law of the Sea attempted to establish a "comprehensive regime of law and order in the world's oceans and seas"; it establishes rules governing all uses of the oceans and their resources. This page gives the text of the Law of the Sea and its amendments, plus links to related information on marine resources, environments, and organizations. *http://www.un.org/Depts/los/index.htm*

Labs and Activities

Southern Illinois University. *Rivers Project Curriculum Units*. White Plains, NY: Dale Seymour Publications, 1998.
This is a series of six river-based units developed by teachers. Units for chemistry, biology, Earth science, geography, language arts, and mathematics use field investigations at local rivers and streams. Available through the publisher at 1-800-237-0338.

Edelstein, Karen, Nancy Trautmann, and Marianne Krasny. *Watershed Science for Educators*. Ithaca, NY: Cornell University, 1998.
This multidisciplinary program demonstrates how to assess physical and biological attributes of local watersheds, including measuring stream depth and velocity, calculating discharges, and sampling aquatic invertebrate life.

Mattingly, Rosanna. "Shedding Light on Watersheds." *The Science Teacher*, November 1991, pp. 46-51.
In this article Mattingly provides instructions for a student activity that includes determining the stream order, pattern, and relief of watersheds.

Kelter, Paul B. et al. "A Discussion of Water Pollution in the United States and Mexico; with High School Laboratory Activities for Analysis of Lead, Atrazine, and Nitrate." *Journal of Chemical Education*, December 1997, pp. 1413-1421.
The authors describe methods suitable for a high school laboratory for analyzing levels of lead, atrazine, and nitrate in samples from local water sources. Available at
http://jchemed.chem.wisc.edu/Journal/Issues/1997/Dec/abs1413.html

Stein, Scott. "Stream Guides." *The Science Teacher*, December 1997, pp. 47-49.
Stein describes a field activity in which students collect water quality data and use it to track a creek's overall health over a period of time.

Myers, Richard L. "Identifying Bottled Water: A Problem Solving Exercise in Chemical Identification." *Journal of Chemical Education*, December 1998, pp. 1585-1587.
The author describes a lab activity in which students analyze the composition of common brands of bottled water (with the labels removed), including characteristics such as hardness, pH level, alkalinity, and presence of metals and nutrients. Students then try to identify each brand by matching the bottle with the chemical characteristics of the water as reported on each brand's label. Available at
http://jchemed.chem.wisc.edu/Journal/Issues/1998/Dec/abs1585.html

Johnson, Robyn, Scott Holman, and Dan Holmquist. *Water Quality with Computers*. Portland, OR: Vernier Software, 1999.
This manual provides instructions for sixteen water quality tests that can be conducted using Vernier sensors with *Logger Pro*. A CD-ROM with experiment files is included with manual.

Johnson, Robyn, Scott Holman, and Dan Holmquist. *Water Quality with CBL*. Portland, OR: Vernier Software, 1999.
This book provides instructions for sixteen water tests that can be done using Vernier sensors with the Texas Instruments Calculator Based Lab (CBL) System and graphing calculators. A CD-ROM with experiment files is included with the manual.

Additional Resources

Herbert, Frank. *Dune*. New York: Ace Books, 1999.
This classic science fiction novel, winner of the Hugo and Nebula Awards, vividly depicts the harsh existence on a desert planet where water is more precious than almost any other substance.

Consortium for Collaborative Experiments: A Bibliography of Books on Rivers of America
This site contains a bibliography of the "Rivers of America" series, published by Farrar & Rinehart beginning in 1935. This series was intended to trace the history and folkways of the United States through its great rivers. There were sixty-five volumes written for this series, covering most of the major rivers in the United States, written by some of best writers in America, and each is full of the stories of the land and the people around the great (and small) rivers. Many of these books are out of print, but may be found in a public library. *http://www.ash.udel.edu/ash/teacher/collab/riversBib.html*
This Library of Congress site discusses the project:
http://lcweb.loc.gov/loc/lcib/970609/text/rivers.html

Powell, John Wesley. *Exploration of the Colorado River of the West and Its Tributaries*. Washington, DC: Government Printing Office. Reprinted as *The Exploration of the Colorado and Its Canyons*. New York: Penguin Books, 1987.
This is Powell's own account of his trips down the Colorado River. The full text of this report, and other writings by Powell, can be found at the University of North Texas' online archive at *http://www.library.unt.edu/gpo/powell/publications.htm*

CHAPTER 3 Biodiversity

Background

Human populations compete with populations of other species for ecosystem resources. Humans, however, have competitive advantages; humans use technology to expand use of Earth's energy and other natural resources. Since humans began to use tools for clearing land to harness fires for heat and other purposes, and to develop weapons for hunting, they have had a measurable impact on ecosystems and habitat. With increased population growth and increased pressure on habitat, concerns have arisen about threats to biological diversity, or biodiversity.

The term "biodiversity" is of relatively recent origin. It was coined during the National Forum on BioDiversity held in 1986 under the auspices of the National Academy of Sciences and the Smithsonian Institution. Biological diversity, or biodiversity, exists on several different levels: the biodiversity of genes within a species, the diversity of individual members of a population within a species, the diversity of species, and the diversity of ecosystems around the world. Of the vast diversity of biological life, from icy arctic lands to the bottom of the ocean, scientists have studied only a small fraction.

> Years before Darwin's visit, Herman Melville described the Galapagos Islands as a desolate fearful place in his work, "the Encantadas or Enchanted Isles":
>
> The Encantadas (the Galapagos Islands) refuse to harbor even outcasts of the beasts. Man and wolf alike disown them. Little but reptile life is here found: tortoise lizards, immense spiders, snakes, and that strangest anomaly of outlandish nature, the aquano (iguana).
>
> ★Herman Melville, **Billy Budd, Sailor, and Other Stories**★

What are the issues?

The importance of preserving Earth's stock of life forms is defended on many grounds: scientific, ethical, aesthetic, and practical. The potential loss of majestic species such as the Sumatran tiger or the Asian panda, so-called "charismatic megafauna," is often highlighted to raise awareness of human threats to endangered species, but ecologists argue that the loss of less-heralded plants and organisms is a concern, since the ecosystem services that these organisms provide is, in many cases, not well understood. As eminent biologist Sir Robert May has said, although there are many organizations for the protection of bird or animal species, there is "no corresponding society to express sympathy for nematodes."

The major threat to biodiversity stems from land use. About 1 to 2 percent of the Earth's surface is devoted to urban use, but modifications of surface area are more extensive. Land acquisition for agriculture and forest products has intensified in many developing countries, which are experiencing rapid population growth. Introduction of species into new areas, either by accident or intention, is a second major cause of threats to species. In the 1890s, a New York City collector introduced a small population of starlings into the city, because he wanted to establish in this country all the birds named by Shakespeare. The starling population flourished, but other populations of songbirds have been diminished. Pollutants and toxins resulting from human activities have also affected ecosystems.

What are the uncertainties?

It has been said that more is known about distant stars in the universe than about the organisms with which we share the planet Earth. It is not known with certainty how many species there are on Earth, or why some areas (the tropics) contain more variety of species than others. Approximately 1.4 million different species have been identified, but this number is uncertain because there is no central registry of species. Most of these individual species are cataloged in museums and research institutions throughout the world and there is some degree of duplication in the records. Certain species, such as mammals and birds, are well known and thoroughly studied; these species, however, make up a tiny fraction of the world's biological diversity. Various estimates have been made of the total number of species by extrapolation, but these estimates range widely and are subject to vigorous debate and revision.

There are varying opinions about the best approach to protecting biodiversity. In past decades, conservation efforts (and funding) have been focused disproportionately on individual endangered species, particularly mammals, but this is costly and concerns have been raised about the adequacy of a species-by-species approach. To provide protection, wildlife preserves and wilderness areas have been set aside in some countries. However, this approach has been criticized as insufficient or ineffective, particularly in developing countries in which the population is heavily dependent upon natural resources for survival. Ecotourism and bioprospecting (deriving medicinal and other products from natural resources) are strategies to help these populations extract products and value from existing forest, but questions have been raised about the effectiveness of these efforts.

What are the risks and tradeoffs?

Various estimates have been made of the value provided by biological diversity, although quantifying these benefits is subject to debate. The importance of soil microorganisms, vegetation, and forests in cycling essential nutrients, of fisheries as a source of protein, and of genetic diversity in plant seeds is well understood. There is also a question of a loss of beauty, of majestic landscapes, and of animals and plants that are unique. Inevitably, though, there are tradeoffs. Establishing preserve areas can pose significant hardship for people in undeveloped areas who depend on the land for sustenance. Funds available for conservation are limited and choices have to be made: Which species and which areas should be protected, and who bears the costs?

BASIC CONCEPTS

An upper-level biology or ecology textbook may be useful for explanations of concepts such as the flow of nutrients and energy through ecosystems, and the meaning of terms such as species, ecosystems, and biomes. Ecology is a highly quantitative field; but there are several introductory level texts and books that will help with a basic understanding. There are also a number of books noted below that have been written for a popular audience and may be easily found in a bookstore or public library.

Botkin, Daniel B., and Edward A. Keller. *Environmental Science: Earth as a Living Planet*. New York: John Wiley & Sons, Inc., 2000.
This introductory college-level textbook includes basic explanations of important ecological concepts such as ecosystem dynamics, biomes, and the relationship between stability and diversity. The text includes helpful equations for calculating production, biomass, and energy flow (pp. 145-146).

Smith, Robert, and Thomas Smith. *Elements of Ecology.* Menlo Park, CA: Addison Wesley Longman, Inc., 1998.
This ecology textbook provides an introduction to basic ecology concepts, including population dynamics, and nutrient and energy flow through ecosystems.

Golley, Frank B. *A Primer for Environmental Literacy.* New Haven: Yale University Press, 1998.
This relatively succinct and inexpensive book provides an excellent introduction to Earth's systems and how they interrelate. Golley presents a clear explanation of the flows of energy and matter through Earth's systems, and of concepts such as speciation, mutualism, competition, and niche.

Ricklefs, Robert E. *Economy of Nature.* New York: W.H. Freeman and Company, 1997.
This is a college-level textbook that provides clear explanations of basic ecology concepts.

Dobson, Andrew P. *Conservation and Biodiversity.* New York: Scientific American Library, 1996.
This edition, by noted biologist Dobson, provides a simply presented overview of issues related to conservation of biodiversity.

Bailey, Robert. *Ecosystem Geography.* New York: Springer-Verlag, 1996.
This reference book delineates the ecosystems in the United States with ecological maps and descriptions of the biotic and abiotic factors that are used to classify ecosystems and determine their boundaries.

The Ecological Society of America: *Issues in Ecology*
This is an online, peer-reviewed journal written by scientists, which includes articles that discuss ecological research in an easy-to-understand manner. Text versions and *.pdf* files are available. *http://esa.sdsc.edu/issues.htm*

Kareiva, Peter M. *Exploring Ecology and its Applications: Readings from American Scientist*. Sunderland, MA: Sinauer Associates, Inc., 1998.
This collection of articles reports on research in the field of ecology, on issues that include the role of species interactions, island biogeography theories, conservation biology, and introduced species. Ordering information is available at
http://www.sinauer.com

Wilson, Edward O. *The Diversity of Life*. New York: W.W. Norton & Co., 1992.
With fascinating stories and rich detail, biologist Edward O. Wilson surveys the origin and role of species diversity and discusses areas of ongoing research in this field.

Reaka-Kudla, Marjorie, Don Wilson, and Edward Wilson, eds. *Biodiversity II: Understanding and Protecting Our Biological Resources*. Washington, DC: Joseph Henry Press, 1997.
This collection of papers by specialists in the field includes discussions of conservation measures, how to classify species rapidly, biotechnology, and ecological restoration.

Bowler, Peter J. *The Environmental Sciences*. New York: W. W. Norton & Co., 1992.
The author presents a history of the field of ecology from its beginnings as a specialization within the field of biology. He notes that the term *oecology* (from the Greek *oikos*, meaning household) was first used by Ernst Haeckel in 1866 to refer to study of the interactions between living organisms and the external world.

What is a Species and How Many are There?

The species is the fundamental unit by which biodiversity is enumerated. Many pages of scientific journals have been devoted to estimating how many species there are on the Earth, and how many there have ever been. A species is the fundamental unit for practical reasons, even though biologists disagree what a species actually is. A species is commonly defined as a population of organisms that potentially can interbreed and reproduce. This definition has limitations, though, and is less useful for asexual microorganisms that exchange genes freely. There are also questions about how different one population must be from another before it is categorized as a new species.

In all cultures, taxonomic classification means survival. The beginning of wisdom, as the Chinese say, is calling things by their right names.

★Edward O. Wilson, **The Diversity of Life**★

By studying fossil records, scientists know more about how many species have ever existed on Earth than they know about how many species currently exist. There is no agreement on this and estimates vary by an order of magnitude. Estimates have been made based on relative sizes of species (smaller ones tend to be more numerous) or by extrapolating from numbers of species that have already been identified. There are many uncertainties, however, in making extrapolations from known species; for example, the animal kingdom represents more than 70 percent of all species that have been identified, but in terms of biomass, animals represent the smallest fraction of all species, far outnumbered by invertebrates.[4] As biologist J. B. S. Haldane said, God seems to have "an inordinate fondness for beetles." And scientists are only beginning to study the remarkable diversity among microorganisms. Only recently scientists have discovered bacteria in hot springs in Yellowstone Park that are unlike any other previously identified.

Gould, Stephen Jay. "What is a Species?" *Discover*, December 1992.
Gould addresses the argument that, since organisms continue to evolve and change, species are arbitrary classifications. He concludes that species is still a useful and necessary concept. Available at *http://www.discover.com/archive/*

May, Robert. "How Many Species Inhabit the Earth?" *Scientific American*, October 1992, pp. 42-48.
Despite over 250 years of systematic research, estimates of the total number of species vary greatly. This article examines the various estimates that have been made, based on factors such as structure of food webs, the relative abundance of species, and their physical size.

Stork, Nigel. "Measuring Global Biodiversity and Its Decline." In *Biodiversity II: Understanding and Protecting Our Biological Resources*. Edited by Marjorie Reaka-Kudla, Don Wilson, and Edward O. Wilson. Washington, DC: Joseph Henry Press, 1997.
The author reviews how much is currently known about biological diversity. Although approximately 1.4 to 1.6 million species have been described, no one knows the precise figure because there is no central registry, and a certain number of the described species are duplicates. Although scientists pursue the question of how many total species there are, Stork notes that not much is known about many of the species that have already been identified.

National Biological Information Infrastructure
This is a collection site for accessing biological data and information from both government agencies and private sector organizations. The site includes links to databases and information on systematics (classification of species). *http://www.nbii.gov/*

Solbrig, Otto. "The Origin and Function of Biodiversity." *Environment*, June 1991, pp. 16-20, 34-38.
This article explains how diversity occurs between and among species, and discusses gene mutation, speciation, resource limitations, and natural selection, in addition to discussing the role of biodiversity within ecosystems.

Morell, Virginia. "On the Origin of (Amazonian) Species." *Discover*, April 1997, pp. 56-64.
The article follows several researchers in the Amazon as they attempt to determine how speciation (the separation of two distinct populations into separate species) occurs among organisms. Among other ideas, the researchers are testing Alfred Wallace's theory that when two populations of a species become separated by rivers or other natural barriers, speciation occurs. Available at *http://www.discover.com/archive/*

Coletta, John, and Erik Munson. "Taxonomy, Ideology and the Structure of the Natural History Field Guide." *The American Biology Teacher*, November/December 1993, pp. 456-462.
The authors note that the traditional species identification schemes found in field guides should not be confused with the rigorous process known as taxonomy. The former uses deductive reasoning to place investigated species into existing classifications, while the latter uses inductive reasoning to group populations and taxa.

ECOSYSTEMS

An ecosystem is defined as a functional unit that results from the interactions of living and non-living things and the flows of matter and energy through, in, and out of the ecological system. Ecosystems can be considered on various scales: from large, such as a temperate forest or a prairie ecosystem, to very small, such as a fallen tree log. There are a number of ecosystems that are just beginning to be explored. In recent years, scientists have found life in areas that were previously believed to be incapable of sustaining life, including volcanoes and steaming hot thermal vents in the deep sea. The basic ecology textbooks listed under Basic Concepts include explanations of ecosystems and their functions. More information on specific ecosystems, such as the Everglades, can be found at *http://www.enviroliteracy.org*

Diamond, Jared, and Ted Case, eds. *Community Ecology.* New York: Harper & Row, 1986.
This is an advanced textbook that covers ecosystem concepts in greater detail.

U.S. Forest Service: Description of the Ecoregions of the United States
This agency site provides summary information about vegetation, soil, flora, and fauna in the various ecoregions of the United States.
http://www.fs.fed.us/land/ecosysmgmt/ecoreg1_home.html

de Blij, H.J., and P.O. Miller. *Physical Geography of the Global Environment.* New York: John Wiley & Sons, Inc., 1996.
This textbook is a useful reference for studying the characteristics of various biomes and ecoregions. It may be available in a public or university library.

Groffman, Peter, and Patrick Bohlen. "Soil and Sediment Biodiversity." *BioScience* 49 (February 1999): 139-148.
The authors describe the relationship between biodiversity of organisms in soil and soil quality. They discuss the role of bacteria and of earthworms, the "charismatic megafauna of the soil," to show that changes in populations of these organisms can alter the cycling of carbon and nitrogen in soil ecosystems. This journal should be available through an electronic database in a public library.

Grassle, Frederick. "Deep-Sea Benthic Biodiversity." *BioScience* 41 (July/August 1991): 464-469.
The author discusses the diversity of organisms that have been discovered in recent years in the dark waters near the ocean floor with the aid of advanced submersible exploration vehicles.

Morell, Virginia. "Life on a Grain of Sand." *Discover*, April 1995, pp. 78-86.
The author explores the complex ecosystem that lies beneath the sands of a Florida beach. She explains the role of microoganisms that consume and recycle the nutrients of the dead fish and seaweed that wash up on shore. Available at
http://www.discover.com/archive/

Jackson, Jeremy. "Adaptation and Diversity of Reef Corals." *BioScience* 41 (July/August 1991): 475-482.
This article examines the characteristics of reef coral ecosystems and reviews what is known about patterns of diversity of coral species.

Wilkinson, Clive, ed. *Status of Coral Reefs of the World: 1998.* Australian Institute of Marine Science, 1998.
This report assesses the health of coral reefs around the world and the various natural and human-related factors thought to be associated with coral bleaching and other damage to these ecosystems. Ordering information and an executive summary of this report are available at *http://www.aims.gov.au/pages/search/search-corals.html*

Davidson, Osha Gray. *The Enchanted Braid.* New York: John Wiley & Sons, Inc., 1998.
Davidson, a journalist, follows along with scientists as they study the "amazing beauty and incredible richness of life" of coral reefs and provides an entertaining natural history of coral reefs and coastal ecosystems.

Krajick, Kevin. "The Secret Life of Backyard Trees." *Discover,* November 1995, pp. 92-98.
Krajick discusses the vital functions performed by specific plant and animal species in forest canopies. Conservationists are studying the habitats of species like the marbled murrelet to understand how to provide protection of their habitat. Available at *http://www.discover.com/archive/*

Rinker, Bruce, Margaret Lowman, and Mark Moffett. "Africa from the Treetops." *The American Biology Teacher,* October 1995, pp. 393-400.
The authors describe their participation in a treetop ecology expedition above the Cameroon rainforest. Researchers were able to access the forest canopy by lowering a large raft from an inflatable airship.

Stability and Diversity

One continuing area of scientific investigation is the relationship between biological diversity and stability. In the early years of ecological studies, scientists speculated that undisturbed communities would reach an equilibrium or "climax" condition. However, ecologists long ago discarded the notion that there is a "balance of nature." In his 1942 book, *Voles, Mice, and Lemmings*, British biologist Charles Eton examined the explosive fluctuations in animal populations from year to year and concluded, "The 'balance of nature' does not exist." It is now understood that ecosystems are subject to many disturbances in addition to human activity, including fires, droughts, or excess rainfall, and that equilibriums are dynamic, changing over time in response to many different factors. The relationship between the diversity of the species in an area and its stability is a matter of uncertainty, and scientific studies have had mixed results.

Pimm, Stuart. "The Complexity and Stability of Ecosystems." ***Nature*** **307 (January 26, 1984): 321-326.**
The author reviews a number of ecosystem studies and questions the theory that increased diversity of species in an ecosystem contributes to its stability and resilience against disturbances.

Baskin, Yvonne. "Ecosystem Function of Biodiversity." ***BioScience*** **44 (November 1994): 657-660.**
Theories concerning the role of biodiversity generally fall under two categories. The first maintains that the loss of any single species to ecosystem functioning is potentially devastating. The other holds that ecosystems comprise more species than necessary to perform each group's role. Recent studies indicate that, although higher diversity levels enhance ecosystem functional diversity, there is a middle ground between the two theories.

Chapin, F. Stuart, et al. "Ecosystem Consequences of Changing Biodiversity: Experimental Evidence and a Research Agenda for the Future." ***BioScience*** **48 (January 1998): 45-52.**
This article examines the relationship between the diversity of plant species in an area and the area's net primary productivity, that is, the energy remaining in an ecosystem after respiration that is stored as organic matter. The study demonstrates that diversity of plant species has an effect on productivity and soil nitrogen, but the underlying mechanisms are not well understood.

Botkin, Daniel. ***Discordant Harmonies.*** **Oxford: Oxford University Press, 1999.**
Illustrated by current and historical case studies, biologist Daniel Botkin reviews how scientists' understanding of nature has changed. Nature undisturbed by human activity is a dynamic, not a static equilibrium. Conservation strategies, he argues, should be based on an understanding of interactions between man and nature.

Threatened Species

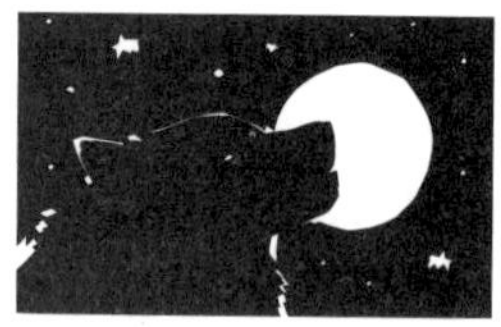

Species are not evenly distributed across the Earth. One study estimates that 70 percent of the world's total species occur in only twelve countries: Australia, Brazil, China, Columbia, Ecuador, India, Indonesia, Madagascar, Malaysia, Mexico, Peru, and Zaire.[5] Unfortunately, these are all countries in which land is rapidly being cleared for use by growing populations. Resources are inevitably scarce, though, so scientists argue for setting priorities for research and conservation efforts. Areas of high priority are sometimes referred to as "hot spots," a term used by biologist Norman Meyers to indicate areas in which there is a high number of endemic species that are found nowhere else and threatened with extinction. It is not known with certainty why tropical areas have more endemic species than other ecosystems, but it is theorized that the constant physical environment of the tropical rain forest has permitted species to evolve a high degree of specialization. This high degree of specialization may make these species more susceptible to human disturbances; for example, if an insect or animal is highly dependent on a specific fruit or tree, loss of that species could have critical effects.[6]

International Union for Conservation of Nature and Natural Resources: The Red List
This is the master list of endangered species around the world.
http://www.iucn.org/themes/ssc/siteindx.htm#table1

World Conservation Monitoring Centre
This is an independent non-profit organization established by the World Conservation Union, the World Wide Fund for Nature, and the United Nations Environment Programme. The Centre site provides a database of endangered species, country profiles, maps, and statistics on marine and forest resources. *http://www.wcmc.org.uk/*

Hayhurst, Chris. "Life after Death: The Rediscovery of 'Extinct' Plants and Animals." ***E*****, November/December 1999.**
Sometimes species long thought to be extinct are rediscovered. Hayhurst reports on biologists who are searching for species believed to be extinct and their success in finding several species of snails, plants, and insects. In one attempt to use "Jurassic Park technology," scientists in New Zealand have made plans to clone the *Huia*, an indigenous bird that was believed to have disappeared in the 1920s. Available at
http://www.emagazine.com/november-december_1999/1199curr_lifeafterdeath.html

Endangered Species: United States

U.S. Fish and Wildlife Service: Endangered Species
This U.S. government agency provides an endangered species database that can be searched by species or state. *http://www.fws.gov/r9endspp/endspp.html*

U.S. Geological Survey Biological Resources Division: ***The Status and Trends of the Nation's Biological Resources*****.**
This report presents the results of the first large-scale assessment of the status of the nation's biological resources – plants, animals, and ecosystems. Among other topics, it includes assessments of major ecosystems in the United States, threats to particular species, and information on land use and water use. *http://biology.usgs.gov/s+t/SNT/*

Wilcove, David et al. "Quantifying Threats to Imperiled Species in the United States." ***BioScience*** **48 (August 1998): 607-615.**
Habitat loss is the single greatest threat to biodiversity in the United States. This article reviews quantitative data relating to threats to plants and animals in this country.

Dobson, A. et al. "Geographic Distribution of Endangered Species in the United States." ***Science*** **275 (January 24, 1997): 550-553.**
The author discusses the "hot spots" of threatened species in the United States, such as Hawaii and Southern California.

Wright, Karen. "Pupfish in Peril." ***Discover*****, July 1999, pp. 42-44.**
In a bottomless limestone cave filled with very hot, very salty, very stagnant water in a deep chasm in the Mojave Desert survives the endangered Devil's Hole pupfish. The pupfish was the subject of a 1976 Supreme Court case that pitted farmers, who relied on the underground water sources that feed the cave, against federal agency efforts to maintain the water levels in the tiny ecosystem. This article describes the risky and difficult semiannual cave-dive by biologists to count the pupfish population.

Mladenoff, David et al. "Causes and Implications of Species Restoration in Altered Ecosystems." *BioScience* 47 (January 1997): 21-31.
The gray wolf was one of the first species listed as endangered under the 1973 Endangered Species Act. Since that time, the wolf population has grown in the areas around the Great Lakes, in large part through another human-caused effect: the growth in deer populations. The authors discuss the "conservation conundrum" that the growing wolf population presents. Wolves are flourishing near human populations and pose threats; large populations of deer that sustain the wolves have destructive impacts on other forest species.

INTRODUCED SPECIES

After loss of habitat, introduced species is the second greatest threat to native plants and animals species. Plants and animals have been brought to new areas intentionally for many reasons. One of the most well known examples is the introduction of rabbits into Australia by a farmer who wanted to hunt rabbits. With no predators, the rabbits multiplied rapidly, causing extensive damage to croplands. Other predators are introduced unintentionally by "hitchhiking." The zebra mussel was accidentally introduced into the Great Lakes by ships emptying ballast water that contained the organisms. Not all introduced species cause a threat to native plants and animals, but it is difficult to predict which ones will become pests. The following resources discuss some of these troublesome visitors.

Pimentel, David et al. "Environmental and Economic Costs Associated with Non-indigenous Species in the United States." June 12, 1999.
This study by Cornell University scientists estimates that the approximately 50,000 non-indigenous species in the United States cause more than $138 billion per year in damages. About 42 percent of the species on the Threatened or Endangered species lists are at risk primarily because of non-indigenous species. Not all introduced species are threats. Introduced species, such as corn, wheat, rice, and other food crops, and cattle, poultry, and other livestock, now provide more than 98 percent of foods in the United States. Available at *http://www.news.cornell.edu/releases/Jan99/species_costs.html*

Hall, Allan. "Costly Interlopers." *Scientific American*, February 15, 1999.
This *Scientific American* article describes the results of the above Cornell University study and provides links to other resources on this topic. Available at
http://www.sciam.com/explorations/1999/021599animals/index.html

Simberloff, Daniel. "The Impact of Introduced Species in the United States." *Consequences*, October 1996.
This article describes how many introduced species, such as the boll weevil, European gypsy moth, and feral hog, migrated into their new environments and how the indigenous ecosystem was affected. Available at
http://www.gcrio.org/CONSEQUENCES/vol2no2/article2.html

Nature Conservancy: America's Least Wanted
This conservation organization provides an extensive resource library on the most troublesome alien invaders. *http://consci.tnc.org/library/pubs/dd/*

U.S. Geological Survey: Non-indigenous Aquatic Species
This U.S. government agency provides scientific reports, data, and general information on aquatic invaders such as zebra mussels. There is also a link to the Non-indigenous Aquatic Nuisance Prevention and Control Act of 1990. *http://nas.er.usgs.gov/*

Simberloff, Daniel, Don Schmitz, and Tom Brown, eds. *Strangers in Paradise: Impact and Management of Non-indigenous Species in Florida*. Washington, DC: Island Press, 1997.
Florida and Hawaii are the U.S. states that are most affected by introduced species, because they are both partly tropical, surrounded by water, and have large areas of land that have been converted to human-dominated habitats. Tropical pets let loose in these states, for example, can flourish and establish populations, which would be impossible in states that experience freezes. This collection of articles discusses how species are introduced, the impact that they have had, and Florida's efforts to mitigate damage from unwanted invaders.

Malakoff, David. "Fighting Fire with Fire." *Science* 285 (September 17, 1999): 1841-143.
Despite some successes with biocontrol, in a few cases species introduced to control weeds and insects have done more harm than good. This article discusses the problems that have occurred when introduced predators have disrupted ecosystems by attacking or displacing non-targeted species.

Hamilton, Gary. "When Good Bugs Turn Bad." *New Scientist* 165 (January 15, 2000): 30-33.
Importing natural predators to control weeds and pests can be cheaper than using chemical pesticides but biologists are concerned about potential threats these introduced species impose on native species.

CONSERVATION

Conservation has a long history in the United States. In the late 19th century Congress established several national parks, including Yellowstone and Yosemite. From the beginnings of the conservation movement, disagreements emerged between two distinct approaches: conservation for the purpose of scientifically managing natural resources for human use, such as for timber, hunting, and fishing; and preservation of wilderness areas protected from human activities. These two sometimes conflicting goals continue to guide differing approaches to conservation today. In developing countries, in which it is thought that most of the world's biodiversity lies, efforts to preserve large tracts of land have proven costly and sometimes ineffective; the growing populations depend heavily on natural resources for survival. In the United States, conflicts arise when conservation programs place heavy financial burdens on those who own or make their living from the land. Much of the Earth's terrain is under human management; increasingly, conservation strategies aim at balancing human needs for natural resources with protection of biological diversity.

Biodiversity Conservation Information System
This "megadatabase" of biodiversity information is sponsored by international conservation groups to provide access to current information about biodiversity and conservation. *http://www.biodiversity.org/*

Green, Michael J. B., and James Paine. "State of the World's Protected Areas at the End of the Twentieth Century." World Conservation Monitoring Centre.
This paper presented at a 1997 IUCN World Commission on Protected Areas Symposium summarizes the status of areas around the world that are protected in reserves and parks, by geographic location, biome, and type of management. The report is available on the World Conservation Monitoring Centre site (*http://www.wcmc.org.uk/protected_areas/albany.pdf*). The Centre site also includes a database that details the amount and type of land in reserves, parks, and other protected areas by country.

National Council for Science and the Environment: Biodiversity & Ecology
This site has a complete listing of the Congressional Research Service's reports on biodiversity and ecology. Topics range from updates on such animals as the Mexican wolf and the manatee to current major global treaties. ***http://www.cnie.org/nle/crsbiodv.html***

Sarkar, Sahotra. "Wilderness Preservation and Biodiversity Conservation: Keeping Divergent Goals Distinct." ***BioScience*** **49 (May 1999): 405-412.**
The author argues that preservation of wilderness areas is a European idea that can be traced to 18th century Romanticism and that creation of wilderness reserves in developing countries where people depend on natural resources for survival may not be the best strategy for conserving biodiversity. This journal should be available through electronic databases in public libraries.

Dobson, Andy et al. "Hopes for the Future: Restoration Ecology and Conservation Biology." ***Science*** **277 (July 25, 1997): 515-522.**
The authors describe how restoration techniques such as bioremediation can mitigate loss of habitat by restoring land degraded by agricultural and industrial uses. Some plants, for example, are capable of accumulating heavy metals from soils and can be used for decontamination.

Access Excellence BioForum: Conservation Biology
This education site presents the text of several scientific lectures on topics including the creation of nature reserves and how researchers assess species for preservation. The scientists trace their own experiences in the field examining endemics, "umbrella" species, and "hot spots." ***http://www.accessexcellence.org/BF/bf06/***

Metrick, Andrew, and Martin Weitzman. "Conflicts and Choices in Biodiversity Preservation." ***The Journal of Economic Perspectives*** **12 (Summer 1998): 21-34.**
Metrick and Weitzman compare biodiversity conservation programs to the loading of Noah's Ark. Planners must decide which species to save given finite resources. The authors suggest how factors such as the utility of each species, the diversity of species, the likelihood of survival, and the costs of increasing the survivability of each species might be quantified to guide decisions about conservation priorities. This journal may be difficult to find; it may be found in an electronic database in a public library.

Westman, Walter. "Managing for Biodiversity." ***BioScience*** **40 (January 1990): 26-33.**
Westman provides an overview of conservation policies and legislation and addresses several unresolved questions, including an interesting issue related to management of exotic species. State park agencies in California would like to eradicate eucalyptus trees from park lands because they are an introduced species; however, several native bird species

and the western population of monarch butterflies prefer the trees. Criteria for deciding which exotic species are undesirable are not always clear cut.

Martin, Rowan. "Biological Diversity." In *Earth Report 2000*. Edited by Ronald Bailey. New York: McGraw-Hill, 2000.
The author, developer of Zimbabwe's CAMPFIRE, a community-based wildlife management program, reviews some of the challenges of preserving biodiversity. He suggests that global planning programs by Western scientists are considered "ecoimperialism" by many struggling developing countries and are ineffective. The Convention on Trade in Endangered Species, for example, he argues, renders some species "legally valueless" and removes any incentive to conserve them. Ordering information is available at *http://www.cei.org/books/earthreport2ak.html*

Goklany, Indur. "Saving Habitat and Conserving Biodiversity on a Crowded Planet." *BioScience* 48 (November 1998): 941-953.
The author assesses the net impact technological progress has had on biodiversity. Although technological change and economic progress have permitted more acres of land to be converted for agriculture and other uses, without recent advances in technology, far more land would have to be converted to grow food. As societies become wealthier, more resources are available for environmental cleanup and more income is derived from knowledge and service sectors, which are less resource-intensive than industry and agriculture.

American National Parks

Huth, Hans. *Nature and the American: Three Centuries of Changing Attitudes*. Berkeley, CA: University of California Press, 1957.
This book traces the American conservation movement from the colonial era to the present, including a history of public parks and the Roosevelt conservation era. This book should be available in a public library.

Sellars, Richard W. *Preserving Nature in National Parks*. New Haven: Yale University Press, 1997.
Sellars, a National Park Service historian, provides a detailed account of how the American national parks system was created and how it has evolved. He describes the tensions that have arisen as the Park Service tries to reconcile sometimes competing management goals including promoting recreation and use of the parks while preserving natural resources.

Library of Congress: Evolution of the Conservation Movement, 1850-1920
This online documentary history of the conservation movement includes articles by many of the leaders of the American conservation movement, including Muir, Pinchot, and Roosevelt, in addition to the original legislation establishing Yellowstone and Yosemite national parks. *http://lcweb2.loc.gov/ammem/amrvhtml/conshome.html*

U.S. National Park Service
The National Park Service's site includes extensive information about each national park, including its history, geology, and natural resources. *http://www.nps.gov*

Banking Biodiversity

World Resources Institute: Ex Situ Conservation
Zoos, botanical gardens, and germplasm banks have undertaken conservation projects. This site describes what each does, discusses the pros and cons of their use, and lists a number of links for each type of institution.
http://www.wri.org/wri/biodiv/ex-situ.html#zoos related sites

Wildt, David et al. "Genome Resource Banks." ***BioScience*** **47 (November 1997): 689-698.**
The organized banking of biomaterials such as blood products and embryos has long been used in agricultural breeding. The authors of this article suggest that since large undisturbed wildlife refuges have become impractical in some parts of the world, genome banks should be used to preserve the genes of endangered species.

Value of Ecosystem Services

The need to conserve biological resources has been justified by the actual and potential value that those resources provide for humans. Ecosystem services include erosion control, climate regulation, nutrient cycling, food and forest and other valuable products. Aspirin, several important cancer drugs, and other medicines were first synthesized from plants. Organisms competing with other species for survival have developed an amazing array of chemical compounds to kill or stun prey and protect them from diseases. Many of these compounds have enormous value when adapted for use as pesticides, drugs, and industrial uses. It has been estimated that one fourth of all prescription drugs used in the United States were derived from plants.

The important potential of these genetic resources for medicines, bioremediation, and other uses has raised concern about the need to quickly assess these resources for commercial value. The commercial value of these products, if it outweighs the value to be gained from clearing the land for agriculture or other uses, provides an incentive to preserve forests. Bioprospecting, though, has become a source of controversy between developing nations, in which many of these substances are found, and Western nations, who normally extract, synthesize, and receive the profits from the drugs. Moreover, finding useful products is a slow and expensive process; it is not clear whether this approach provides enough incentive to protect forest areas.

Costanza, Robert et al. "The Value of the World's Ecosystem Services and Natural Capital." ***Nature*** **387 (May 15, 1997): 253-259.**
The authors have estimated that services provided by the Earth's ecosystems, including food, recycling of water and nutrients, pollination, raw materials, and recreation, has an average minimum value of $33 trillion per year. Although it is argued that it is impossible or unwise to place a value on intangibles such as the aesthetic pleasure derived from nature, the authors suggest that valuations are made every day implicitly and that an accurate reckoning of the costs would better inform decisions.

Pimentel, David et al. "Economic and Environmental Benefits of Biodiversity." ***BioScience*** **47 (December 1997): 747-757.**
These authors propose an estimate of the value of ecosystem services provided by biological diversity, including ecotourism, cycling of nutrients, medicines and other products.

Mukerjee, Madhusree. "Sowing Where You Reap." *Scientific American*, May 1996.
This article examines conflicts that arise over intellectual property rights and profits gained from bioprospecting ventures in developing countries. Some developing countries are limiting access to bioprospecters to ensure they receive a share of the profits from pharmaceuticals derived from their native plants and organisms. Available at *http://www.sciam.com/0596issue/0596scicit07.html*

Eisner, Thomas, and Elizabeth Beiring. "Biotic Exploration Fund: Protecting Biodiversity Through Chemical Prospecting." *BioScience* 44 (February 1994): 95-98.
Using the INBio-Merck agreement as an example, the authors present the debate over intellectual property rights to genetic resources found in developing countries.

Kaiser, Jocelyn. "Unique, All-Taxa Survey in Costa Rica 'Self-Destructs.'" *Science* 276 (May 9, 1997): 893.
Although the INBio project in Costa Rica has been viewed as a model for other bioprospecting projects, the survey was cancelled by Costa Rica's National Institute for Biodiversity (INBio) in November 1996. According to this *Science* magazine article, INBio officials believed the program focused more on scientific goals than on the economic benefits that the country needs to protects its forests.

World Resources Institute: A Sample of Companies Active in Plant and Other Natural Product Collection and Screening
This site includes a list of companies involved in bioprospecting and a short list of important drugs derived from plants. *http://www.wri.org/wri/biodiv/b01-bp.html*

Mestel, Rosie. "Drugs from the Sea." *Discover*, March 1999, pp. 70-75.
The sea may be the next frontier for new pharmaceutical products. This article reports on researchers' attempts to retrieve and study sea life for their potential medicinal value. Available at *http://www.discover.com/recent_issue/index.html*

Crystall, Ben. "Monstrous Mucus." *New Scientist* 165 (March 11, 2000): 38-41.
The potential of finding valuable products from the rich variety of life on Earth is difficult to calculate or predict because useful products sometimes come from strange sources. This article reports on research on the hagfish, which deters predators by shooting out great quantities of slime. Researchers have found that this slime, which contains silk-like, thin, very strong fibers, may pave the way for new super-strong materials.

Johns, Timothy. *The Origins of Human Diet and Medicine: Chemical Ecology*. Tucson, AZ: University of Arizona Press, 1996.
The same properties that make some plants inedible as food makes them good candidates for medicines. In this body, originally published under the title, *With Bitter Herbs They Shall Eat*, Johns explains the chemical defenses that plants have evolved to defend themselves from predators and examines how humans have discovered, among the many poisonous and toxic plants, which ones are safe to eat and which ones have medicinal properties. This is not an easy book, but would be useful for further research on this topic. The book can be ordered at *http://www.uapress.arizona.edu/*

Endangered Species Act

The Endangered Species Act, enacted into law in 1973, directs the U.S. Fish and Wildlife Service to maintain a list of species that are either in danger of extinction or likely to become in danger. The agency is empowered to halt actions that threaten species on the list and to work to recover the species so further protection is unnecessary. From the beginning, the Act has been controversial. Property owners suffer financial losses if they are unable to develop land that has been designated a critical habitat for an endangered species. The Act has been criticized for being ineffective, and only a few species have been recovered under the Act. Congressional authorization for spending under the Endangered Species Act expired in 1992, but efforts to reauthorize or amend the Act have been unsuccessful.

To reduce litigation and conflict between private property owners, and state, local, and federal officials over protection of endangered species, the Department of the Interior has made more frequent use of a provision in the Act for establishing habitat conservation plans. These agreements permit property owners to develop property in areas where endangered species are found, in exchange for setting aside acreage elsewhere. Since most habitats for endangered species are on lands owned by private citizens, protection efforts are difficult without their cooperation. Individual owners, not the taxpayers, have to bear the costs (often substantial) for protecting habitat. Although the Department of the Interior has promoted this approach as an alternative to "environmental trainwrecks," property owners, environmental groups, and scientists are divided on the issue.

Mann, Charles, and Mark Plummer. "Is the Endangered Species Act in Danger?" ***Science*** **267 (March 3, 1995): 1256-1258.**
The authors review progress under the Endangered Species Act. Of the 955 species that were listed as endangered or threatened from 1973 to 1994, only 21 species have been "delisted" and most of these were removed from the list because they went extinct, were listed through error, or recovered through measures unrelated to the Endangered Species Act. Because critics disagree about why the Act is ineffective, widely divergent proposals to amend the Act have been introduced.

National Research Council. ***Science and the Endangered Species Act.*** **Washington, DC: National Academy Press, 1995.**
This report reviews issues related to the current endangered species program, discussing the definition of "species," the role of habitat conservation, conservation conflicts between species, recovery planning measures, and the tradeoffs that arise. Available to read or purchase at *http://www.nap.edu*

Kaiser, Jocelyn. "When a Habitat Is Not a Home." ***Science*** **276 (June 13, 1997): 1636-1638.**
This article reviews the controversy over habitat conservation plans and criticisms by ecologists and others concerning the plans.

U.S. Fish and Wildlife Service: Endangered Species Habitat Conservation Planning
This site explains habitat conservation planning and includes a handbook and links to related resources. *http://endangered.fws.gov/hcp/index.html*

LAWYERS, WOLVES, AND THE GOD SQUAD

Shortly after the Endangered Species Act was signed into law, the first major battle between state, local, and federal agencies began. The debate over the Tennessee Tellico Dam project and the snail darter lasted for twelve years and involved state and federal officials, the U.S. Supreme Court, Congress, and the "God Squad," the special committee established under the Act to decide whether a development project should go forward even though it might result in the extinction of a species. This suit and other cases involving the Endangered Species Act provide fascinating case studies of the often conflicting biological, economic, and political interests involved in these issues. Even animals have been given standing to sue in federal court, giving rise to case names such as Mount Graham Red Squirrel v. Yeutter, 930 F.2d 703 (9th Cir. 1991) and Loggerhead Turtle v. County Council, 896 F.Suppl 1170, 1178 (M.D. Fla. 1995).

One issue concerns the fate of wolves in Yellowstone National Park. In 1996, a U.S. District Court ruled that wolves from Canada the U.S. Fish and Wildlife Service had reintroduced into Yellowstone National Park had to be removed. The following resources provide more information on this issue.

The Total Yellowstone Wolf News Page

This site gives wolfpack updates and tracks the political debate over the wolves introduced in Yellowstone National Park. http://www.yellowstone-natl-park.com/wolf.htm

Audubon Society

The Audubon Society, along with the Farm Bureau, sued for removal of the wolves from the park. The Farm Bureau sued on behalf of ranchers who lose their domestic animals to wolves. The Audubon Society provides a brief history of their part in the Yellowstone wolf debate and links to news updates and photographs of wolves in the forest. http://www.audubon.org

National Center for Policy Analysis: Dangers of Wolves in the Wild

This policy group summarizes why farmers are resistant to the reintroduction of wolves. http://www.ncpa.org/pi/enviro/envpd/pdenv139.html

Meltz, Robert. "The Endangered Species Act and Private Property: A Legal Primer." Congressional Research Service Report for Congress, No. 93-346A, March 7, 1993.
This report analyzes controversies concerning the Endangered Species Act, which has not been reauthorized by Congress. Available at *http://www.cnie.org/nle/biodv-8.html*

Stroup, Richard. "Endangered Species Act: Making Innocent Species the Enemy." ***Political Economy Research Center Policy Series*****, Issue No. PS3, April 1995.**
This policy brief argues that the Endangered Species Act creates perverse incentives that punish landowners who try to protect endangered species on their land. Available at *http://www.perc.org/ps3.htm*

Corn, Lynne. "The Listing of a Species: Legal Definition and Biological Realities." Congressional Research Service Report for Congress, No. 92-944ENR, December 15, 1992.
Not all species are equal under the Endangered Species Act. Individual vertebrates, such as birds and mammals, are protected on both public and private lands, but plants are not. Invertebrates also receive lesser protection. As this report details, the Act's definition of species, and its distinctions between them, rest on legal definitions not scientific ones. Available at *http://www.cnie.org/nle/biodv-10.html*

Flather, Curtis et al. "Threatened and Endangered Species Geography." ***BioScience*** **48 (May 1998): 365-376.**
Since the 1973 Endangered Species Act, the federal government has employed a species conservation program aimed at protecting the most endangered species on a species-by-species basis. The authors of this article assert that this strategy ignores the dynamics of

ecosystems, and suggest that the protection of geographic "hot spots" would be more effective.

Legislation and Treaties

Flora and Fauna – Biodiversity
This Tufts University site has a comprehensive list of treaties and conventions relating to biodiversity. *http://fletcher.tufts.edu/multi/biodiversity.html*

Endangered Species Act (16 USC sec. 460) (7 USC sec. 136)
The full text of the act governing the conservation of threatened and endangered plants and animals (and their habitats) appears at this university site.
http://www4.law.cornell.edu/uscode/16/ch35.html

Convention on Biological Diversity
The Convention is the international agreement for biological conservation policies. This site includes databases and text of a number of the convention's policy documents.
http://www.biodiv.org/

Fletcher, Susan. "Biological Diversity: Issues Related to the Convention on Biodiversity." Congressional Research Service Report for Congress, No. 95-598 ENR, May 15, 1995.
This report covers the text of the Convention on Biological Diversity, its history and current status (including legislation), and provides background information. Available at
http://www.cnie.org/nle/biodv-2.html

Convention on Trade in Endangered Species (CITES)
This site lists transcripts of the convention and related documents. The site also links to other online biodiversity databases, including the text of a number of biodiversity treaties and a listing of species locations worldwide. *http://www.cites.org/*

Federal Noxious Weed Act
This site gives the text of the act that governs the dissemination of introduced weed species that pose a potential threat to indigenous ecosystems.
http://bluegoose.arw.r9.fws.gov/FICMNEWFiles/FederalNoxiousWeedAct.html

Lacey Act of 1900
This site gives the text of the act that governs illegally taken fish and wildlife and restricts non-native species' entrance into the United States. *http://www.nmfs.gov/*

Labs and Activities

Ecological Society of America: Education Section
This professional association's site includes online instructions for "Experiments to Teach Ecology," including a population ecology model using duckweed and activities to study foraging and flocking behavior.
http://www.science.widener.edu/~grant/esa/edweb.html#vol2

LaBare, Kelly M., R. Lawrence Klotz, and Eric Witherow. "Using Online Databases to Teach Ecological Concepts." *The American Biology Teacher*, February 2000, pp. 124-127.
Ecology studies are often longitudinal studies, assessing change over a period of time, which is difficult to accomplish in a one-year course. The authors demonstrate how a long-term study of an ecosystem can be accomplished by using data obtained through the Internet. The National Science Foundation-funded Long-Term Ecological Research (LTER) program sites post their data on the Internet. The author shows how to download and manipulate the ecological data.

The Microbial World: Winogradsky Columns—Perpetual Life in a Tube
This site demonstrates how to set up a Winogradsky column, a tube that is a complete, self-contained recycling system, deriving energy from light. These columns can be used to demonstrate nutrient cycling by microorganisms in an ecosystem.
http://helios.bto.ed.ac.uk/bto/microbes/winograd.htm

Woodrow Wilson Leadership Institute for Teachers: The Winogradsky Column
This site also includes a basic lab for creating Winogradsky columns.
http://www.woodrow.org/teachers/esi/1999/princeton/projects/microbe/win_col.html

Advanced Arthropod Studies Curriculum
This curriculum, which was developed by students at the University of Illinois at Urbana-Champaign, includes units on collecting arthropods for study using a Berlese funnel (an easy-to-construct collection device), studying arthropod variation and taxonomy, and quantitatively analyzing arthropod populations.
http://www.uni.uiuc.edu/~dstone/advancedarthropodstudies.html

Peers Make and Take: Berlese Funnel
This site provides instructions for constructing a Berlese funnel using simple materials that can be used to collect invertebrates.
http://140.211.62.101/lessons/makentake/berlese1.html

Additional Resources

Quammen, David. *The Song of the Dodo: Island Biogeography in an Age of Extinctions.* New York: Charles Scribner's Sons, 1996.
Nature writer Quammen provides an engaging account of scientists from Darwin to E.O. Wilson who studied the origin, evolution, and extinction of species on islands. The book ranges from scientists' theories about swimming elephants and the grueling aspects of field work, to how the field of conservation biology developed.

Weiner, Jonathan. *The Beak of the Finch.* New York: Vintage Books, 1995.
The author won a Pulitzer Prize for this account of the work of two scientists who have spent more than twenty years studying natural selection in action among Darwin's finches in the Galapagos Islands. The book demonstrates how species are shaped by the flora, fauna, soil, and climate in their environment, and how populations rise and fall over seasons, years, and decades.

DeBlieu, Jan. ***Wind: How the Flow of Air Has Shaped Life, Myth, and the Land*****. Boston: Houghton Mifflin, 1998.**
The author presents a fascinating natural history of wind with discussions of topics such as how winds help plants and insects disperse across land and water, how wind shapes sand dunes and other landforms, and how "ill-winds," such as the *chinook* and the *Santa Ana* are thought to affect humans.

Hudler, George. ***Magical Mushrooms, Mischievous Molds*****. Princeton: Princeton University Press, 1998.**
Hudler combines a discussion of the science of fungi with its role throughout history. Fungi have served as food, poison, drugs, and a myriad of other functions. The substance has played a part in the Salem Witch Trials, the Great Irish Potato Famine, and world wars.

Plotkin, Mark. ***Tales of a Shaman's Apprentice: An Ethnobotanist Searches for New Medicines in the Amazon Rain Forest*****. New York: Penguin Books, 1994.**
This book details research conducted by an ethnobotanist among native healers in the Amazon forest to identify native cures used for arthritis, skin fungi, colds, and other afflictions.

Lembke, Janet. ***Despicable Species: On Cowbirds, Kudzu, Hornworms, and Other Scourges*****. New York: The Lyons Press, 1999.**
This is a natural history of several species that are regarded as pests or nuisances. Lembke, for example, explains how kudzu was imported by a well-meaning horticulturalist and planted widely by the Department of Agriculture during the Dust Bowl years to prevent erosion.

Orlean, Susan. ***The Orchid Thief*****. New York: Random House, 1999.**
Orlean combines an account of a true story about a south Florida plant dealer who was arrested for stealing endangered orchids from a state preserve, with a fascinating history of the obsession with this beautiful and fragile *epiphyte*.

Lewis, Mark. ***Cane Toads: An Unnatural History*****. First Run Features, 1988.**
This sometimes humorous documentary tells the tale of *Bufo Marinus* toads in Australia. The large, poisonous-skinned toads were imported from Hawaii in the mid 1930s to control sugar cane pests. The toads were not successful as biocontrol agents, but they have spread across the Austrialian continent. They are regarded as a nuisance by some, and loved by others. Available from any well-stocked video store.

Knutson, Roger M. ***Fearsome Fauna: A Field Guide to the Creatures that Live in You*****. New York: W.H. Freeman and Company, 1999.**
Biologist E.O. Wilson has said that, "an entire ecosystem can exist in the plumage of a bird." You can, as this book makes clear, find an ecosystem within. Roger M. Knutson, professor emeritus of biology, provides a fascinating and amusing look at the amazing number of parasites that have made the human body their environment. This is an inexpensive book that should be readily available at most large bookstores or public libraries.

The Carnivorous Plant FAQ
There is an incredible variety of life on Earth, including carnivorous plants. This site, by weed scientist Barry Meyers-Rice, explains how plants trap and eat insects, and even rats, birds, and frogs. *http://www.sarracenia.com/faq.html*

Background

> The poison winds, such as the sirocco of the Mediterranean, the brickfielder of Australia, the xlokk of Malta, the simoon of North Africa, and the khamsin of Egypt and Isreal, all drain the energy from the bones... On nights when the Santa Ana blows, (novelist) Raymond Chandler writes, "every booze party ends in a fight. Meek little wives feel the edge of the carving knife and study their husbands' necks."
>
> ★Jan DeBlieu, **Wind: How the Flow of Air Has Shaped Life, Myth, and the Land**★

Air pollution has been a problem for humans at least since the first campfire was brought indoors. Ancient Roman writers complained about the effects of wood smoke on health. And around the year 1300, in perhaps the first act of environmental regulation, England's King Edward II decreed that no coal could be burned in London while Parliament was in session. Modern concern over air pollution became most intense beginning with the economic boom that followed World War II. As manufacturing grew and a growing number of families could afford automobiles, many areas experienced a continually increasing concentration of atmospheric smoke, dust particles, and exhaust fumes. In Donora, Pennsylvania in 1948, and London, England in 1952, episodes of noxious weather-related smog pollution that persisted for days or weeks contributed to numerous fatalities and focused public attention on the need to address this problem. Efforts by state and local governments to improve air quality began in the late 1940s. In 1950, President Truman convened the first national air pollution conference.

The first federal Clean Air Act was passed in 1963, and major improvements in air quality have been registered since then, even as the total number of miles driven by motorists has doubled. According to the Environmental Protection Agency's 1997 report to Congress, emissions of all major air pollutants decreased from 1970 to 1994, with the exception of nitrogen oxides. Emissions of the most dangerous pollutant, lead, decreased by 98 percent, a remarkable success story.

What are the issues?

Smog and soot particles reduce visibility and can cause health problems, particularly for the elderly and those with respiratory problems. These pollutants are primarily a local problem, but can be transported by winds to affect other areas. Sulfur dioxide from coal-burning power plants and nitrogen oxides emitted from cars mix with water vapor to increase the acidity of precipitation, which can affect the chemical makeup of soils and fresh water lakes in certain regions. And with the increasing prevalence of new energy-efficient, airtight homes and offices, indoor air quality has increasingly become a matter of concern, since dust particles and pollutants are less likely to escape outdoors.

Although there has been a significant decline in most major air pollutants over the last two decades, nitrogen oxides have not declined significantly. These emissions are implicated in the production of ozone and acidic precipitation.

What are the uncertainties?

New regulations have recently been promulgated under the Clean Air Act Amendments of 1990 that set lower thresholds for pollutants, including ozone and small particles. However, questions have been raised about whether any health benefits are produced by these lower thresholds. There is no question that at certain levels, ozone and dust particles can cause respiratory distress, particularly among vulnerable populations; however, the biological mechanisms are not well understood, and questions have been raised, for example, about whether it is the size of a particle, its composition, or some other factor that may cause harm.

What are the risks and tradeoffs?

Further reductions in emissions could be costly and difficult to accomplish. The easiest and most cost-effective measures have already been implemented. In many areas, further reductions would require changes in individual behavior, including restrictions on driving, to achieve compliance with revised air quality standards. Increasingly stringent emissions control also raises the costs of vehicles, forcing many motorists to delay purchasing new cars, leaving a higher percentage of older vehicles on the road with emissions control equipment that works less effectively.

Ironically, a number of measures that have been taken to reduce the emissions of pollutants into the atmosphere have contributed to other environmental problems. Tall smoke stacks, for example, were built to alleviate local soot and smoke problems. But because the height of the stacks allows the sulfate smoke to mix with atmospheric water vapor to form acid precipitation, a local air quality problem was transformed into a regional water quality issue. One emerging environmental issue concerns MTBE (methyl tertiary-butyl ether), an oxygenate that was added to gasoline to make it burn cleaner and reduce air emissions. Incidents of groundwater contamination led California to phase out use of MTBE and EPA's scientific advisory panel on this issue has recommended that the MTBE requirement be removed. Many of the approaches taken in the past to reduce pollutants have transformed one type of pollution to another, or from one medium (air) to another (water).

Basic Concepts

Air pollutants originate from both natural and human sources. Natural sources include volcanic eruptions, sea spray, forest and grass fires, and wind-blown dust. Human sources include cars and trucks, power plants and factories, and agriculture. Some pollutants, such as carbon monoxide, may cause harm when directly emitted into the air. These are categorized as primary pollutants. They may undergo chemical change in the air as a result of reaction to naturally-occurring atmospheric chemicals or natural processes such as solar radiation, and in some cases, the secondary pollutants formed may cause more environmental harm than the primary pollutants. Acid rain from sulfur dioxide and ground level ozone from nitrogen oxides are two examples of secondary pollution.

Chemistry textbooks may be helpful for their explanations of the chemical processes related to air quality issues. Earth sciences textbooks would also be useful in describing the composition of the atmosphere and the natural and meteorological processes that affect air quality. Many of the books and articles below may be found in a public or university library.

Botkin, Daniel, and Edward Keller. *Environmental Science: Earth as a Living Planet*. New York: John Wiley & Sons, Inc., 1998, Ch. 22, pp. 466-495.
This introductory environmental science textbook provides a brief overview of major air pollutants, their sources, and health effects.

Jackson, A., and J. Jackson. *Environmental Science: The Natural Environment and Human Impact*. Essex, UK: Longman Publishing, 1996, pp. 298-316.
This more advanced text is useful for its explanations of the chemical makeup of the major pollutants and the chemical reactions that produce major secondary pollutants such as ozone and acid rain.

Boubel, Richard et al. *Fundamentals of Air Pollution*. New York: Academic Press, 1994.
This book is a general, comprehensive introduction for the non-specialist. It explains the chemistry of air pollution and methods of measuring pollutants, the weather conditions that contribute to formation and transport of pollutants, and the engineering problems involved in pollution control.

Godish, Thad. *Air Quality.* Third Edition. New York: Lewis Publishers, 1997.
This is a detailed, moderately technical book covering the chemistry of air pollution and the interactions between emissions and atmospheric processes.

Stanitski, Conrad, ed. *ChemCom: Chemistry in the Community.* Dubuque, IA: Kendall/Hunt Publishing Co., 1996.
This text, a project of the American Chemical Society funded by the National Science Foundation, is intended to provide a basic introduction to chemistry and its impact on society. It is a useful reference book for its explanations of chemical reactions relevant to discussions of acid rain, smog formation, and emissions control technology.

Buell, Phyllis, and James Girard. *Chemistry: An Environmental Perspective*. Upper Saddle River, NJ: Prentice Hall, 1994.
This college textbook may be useful as a reference because it includes detailed explanations of the chemical reactions involved in issues related to air quality (and other topics).

Historical Background

Hong, Sungmin et al. "History of Ancient Copper Smelting Pollution during Roman and Medieval Times Recording in Greenland Ice." *Science* 272 (April 12, 1996): 246-249.
By analyzing ice cores from Greenland, researchers found that emissions from copper smelting were greater during Roman and medieval times than during the period from the Industrial Revolution to the present day.

Goklany, Indur. "The Environmental Transition to Air Quality." *Regulation* 21 (Fall 1998): 36-46.
Goklany argues that improvements in air quality had been made by states before the passage of the first federal Clean Air Act. He argues that a general process leads from economic affluence to investments in environmental quality, and that the states were already following this model by cleaning up their air when the 1970 Clean Air Act made the issue a federal concern.

Urban Air Pollution

Urban areas tend to have the highest concentrations of pollutants from manufacturing and transportation sources. One of the most obvious urban air quality problems is smog, which is formed when emissions from power plants and automobiles react with sunlight or atmospheric compounds. Photochemical or brown smog consists primarily of ozone, which is formed in complex chemical reactions involving nitrogen oxides in the presence of strong sunlight and the absence of winds. Incidents of ground-level ozone are therefore temporary, and occur only during daylight hours, typically on hot summer days.

Industrial or gray smog is a winter phenomenon that occurs when sulfur dioxide emissions react with water vapor. The result is airborne sulfuric acid which may have more severe health effects than those of brown smog. Smog emergencies caused by the buildup of industrial sulfur emissions occurred in London in 1952 and Donora, Pennsylvania, in 1948.

Urban air pollution is highly dependent on meteorological conditions. Wind and geographic factors determine whether emissions are dispersed. Serious air pollution events are often caused or aggravated by a weather condition known as an atmospheric inversion. In an atmospheric inversion, a layer of cool air aloft blocks the normal upward movement of warm air, preventing the dispersion of pollutants from cars and factories. This often happens in urban areas where nearby mountains form a natural barrier to the movement of air. Los Angeles, which has dense automobile traffic, bright sunshine, and mountains to the east that block winds from the ocean, is particularly subject to smog conditions.

Environmental Protection Agency: Ozone: Good Up High, Bad Nearby
This federal government site provides a basic illustrated explanation of the differences between stratospheric and tropospheric ozone.
http://www.epa.gov/oar/oaqps/gooduphigh/

Seinfeld, John H. "Urban Air Pollution: State of the Science." ***Science*** **243 (February 10, 1989): 745-752.**
This article discusses the sources of ozone air pollution. The author notes that current regulations rely almost exclusively on nitrogen oxides emissions control and argues that monitoring of the actual sources of ozone-producing chemicals would be more effective.

Lents, James M., and William J. Kelly. "Clearing the Air in Los Angeles." ***Scientific American*****, October 1993, pp. 32-39.**
The authors discuss the technological and regulatory efforts to curb air pollution in Southern California, where the problem of automobile emissions is compounded by the region's vulnerability to atmospheric inversions.

Kiester, Edward, Jr. "A Darkness in Donora." ***Smithsonian*****, November 1999, pp. 22-24.**
The author interviews residents of Donora, Pennsylvania, who share their memories of the 1948 smog that killed twenty people in this Pennsylvania factory town. Available at
http://www.smithsonianmag.si.edu/smithsonian/issues99/nov99/phenom_nov99.html

Kiester Jr., Edward. "Prophets of Gloom." ***Discover*****, November 1991, pp. 52-56.**
Lichens are rootless plants that obtain all their nutrients from the air, not from the soil, so they can be used to monitor concentrations of pollutants in the air. For example, by

tracking the path of lichen damage, Finnish researchers were able to trace the source of pollutants to a nickel smelter in the former Soviet Union.

Puget Sound Clean Air Agency: Our Central Puget Sound Airshed
This local site describes how air currents and geography contribute to Seattle's air quality. This page provides a good introduction to the phenomenon of an atmospheric inversion. *http://www.psapca.org/airshed.htm*

Woodrow Wilson Leadership Program in Chemistry
This site includes an environmental chemistry activity that demonstrates an atmospheric inversion. *http://www.woodrow.org/teachers/chemistry*

Chaikin, Don. "How It Works: Catalytic Converter." ***Popular Mechanics*****, April 1992, p. 98.**
Chakin describes the design of a catalytic converter and the chemical reactions that occur within it.

ACID PRECIPITATION

Rain is naturally acidic as a result of the reaction of water vapor, nitrogen, and carbon dioxide in the atmosphere. Acidity of rain can be increased through the introduction of sulfur dioxide and nitrogen oxides into the atmosphere. This can occur naturally from sources such as sea spray and volcanic eruptions. The primary human contributions of sulfur dioxide and nitrogen oxides to the atmosphere come from fossil fuel combustion, particularly power plants and automobile exhaust. Increased acidity of precipitation can change the chemistry of soils and fresh water lakes, affecting vegetation and aquatic life, although the effects vary according to local conditions. The term "acid rain" is often used, but "acidic precipitation" is a more accurate term because it can also reach the ground as dry particles, mist, sleet, or snow as well as rain.

U.S. Geological Survey: Online Data and Reports on Acid Rain, Atmospheric Deposition and Precipitation Chemistry
This federal government site houses a database of precipitation data in the U.S. and reports on trends and environmental effects of acidic deposition. The site also provides a primer on how acid deposition is formed. *http://btdqs.usgs.gov/acidrain/*

National Acid Precipitation Assessment Program. ***1990 Integrated Assessment Report.*** **Washington, DC: Government Printing Office, 1991.**
This report was the result of a ten-year study on the causes and effects of acid precipitation. The study found that the effects of acidic precipitation were more modest than originally hypothesized.

National Acid Precipitation Assessment Program Biennial Report to Congress: An Integrated Assessment
The Clean Air Act Amendments of 1990 established an allowance trading program for sulfur dioxide emissions. According to this program, electricity generating plants are assigned allowable levels of emissions. If their emissions are less than their allowed level, they can save ("bank") these credits against future emissions or trade them to other power plants. This site contains the results of an assessment of the effectiveness of this program. *http://www.nnic.noaa.gov/CENR/NAPAP/*

Howells, Gwyneth. ***Acid Rain and Acid Waters.*** **Second Edition. New York: Ellis Horwood, 1995.**
This is a fairly technical but thorough treatment of the major issues surrounding acid deposition. It discusses the sources of sulfur dioxide and nitrogen oxides that lead to acid deposition, its effects, and methods of mitigating the problem.

CHEMystery: Acids and Bases
This educational site for students includes a detailed explanation of acids and bases, which may be helpful in explaining the pH scale and what it measures.
http://library.advanced.org/3659/acidbase/acidsandbases.html

McCormick, John. "Acid Pollution: The International Community's Continuing Struggle." ***Environment,*** **April 1998, pp. 17-20, 41-45.**
The author traces the development of international acid pollution regulations and goals from the first recognition of the problem in the early 1900s in Britain to the present day.

Kerr, Richard, and Jocelyn Kaiser. "Acid Rain Control: Success on the Cheap." ***Science*** **282 (November 6, 1998): 1024-1027.**
Kerr and Kaiser report that the emissions trading program of the Clean Air Act of 1990 has spurred reductions in sulfur dioxide emissions by lowering the cost of reductions.

Air Quality in Developing Nations

Emissions control and energy efficiency technologies have led to significant improvements in air quality in the U.S. in the past two decades. However, rapidly urbanizing and industrializing countries around the world face seriously declining urban air quality. Cities such as Mexico City and Bangkok have rapidly growing populations and increasing numbers of automobiles with high emissions of lead and other pollutants. Indoor pollution is also a serious environmental health problem in many developing nations. The greatest exposures come from by-products of indoor combustion for cooking and heating, which causes respiratory illnesses, especially in women and children, who tend to spend more time indoors on household tasks.

"Air Pollution in the World's Megacities." ***Environment,*** **March 1994, pp. 4-13, 25-37.**
This article summarizes a study of air quality in twenty of the world's largest cities by the Global Environmental Monitoring System, an environmental assessment program of the United Nations.

United Nations Environmental Programme: State of the Environment Reports
This site provides data on countries throughout the world, indicating the environmental problems and policies of each country, including air quality.
http://www.unep.org/unep/soe.htm

World Bank: Urban Air Quality Management
This World Bank site discusses urban air quality problems in developing countries, especially in cities such as Karachi, Seoul, and Santiago. The site also discusses policies to reduce pollution. *http://www.worldbank.org/html/fpd/urban/air_poll/air_poll.htm*

Indoor Air Quality

Improvements in energy efficiency of homes and offices can lead to a deterioration of indoor air quality in poorly ventilated spaces. Better insulation and more tightly sealed windows and doors decrease circulation of fresh air, leading to higher concentrations of pollutants. In the industrialized world, increased incidence of childhood respiratory illnesses such as asthma are thought to be related to increased exposure to indoor pollutants, including dust mites and cockroach droppings. Increased exposure may arise both from more tightly sealed homes and offices and from the longer periods of time young people spend indoors compared to decades past.

Ott, Wayne R., and John W. Roberts. "Everyday Exposure to Toxic Pollutants." ***Scientific American*****, February 1998, pp. 86-91.**
The authors argue that public health efforts relating to pollution should focus more on the greater exposure threat represented by indoor air contaminants than on the relatively temporary exposure to outdoor contaminants. Available at
http://www.sciam.com/1998/0298issue/0298ott.html

Schoen, Deborah. "Fungal Sleuths." ***BioScience*** **48 (January 1998): 9-10.**
This article discusses research on indoor microbes and fungi. Among findings are that all fungi potentially can cause allergic reactions, and that some produce toxins that can have serious health effects.

Occupational Safety and Health Administration: Legionnaires' Disease
Contaminants pose the greatest risk when they are dispersed through heating, air conditioning, and ventilation systems. The Legionnaires' Disease outbreak of 1976, one of the most well known cases of such contamination, caused thirty-four deaths.
http://www.osha-slc.gov/SLTC/legionnairesdisease/index.html

Platts-Mills, Thomas A.E., and Melody C. Carter. "Asthma and Indoor Exposure to Allergens." ***The New England Journal of Medicine*** **336 (May 8, 1997): 1382-1384.**
This article discusses the increase in incidence and severity of childhood asthma in recent years, examining the allergic effects of natural pollutants such as dust mites, cockroach droppings, and pet dander. Available at
http://www.nejm.org/content/1997/0336/0019/1382.asp

American Council for Science and Health. ***Priorities For Health.***
This online periodical by the American Council for Science and Health includes articles by scientists and physicians that address health concerns that are often misreported in the media. Recent issues have included articles on environmental health issues such as asthma, multiple chemical sensitivity, and asbestos. *http://www.prioritiesforhealth.com*

Moore, Cassandra. ***Haunted Housing.*** **Washington, DC: Cato Institute, 1997.**
Moore examines the debates surrounding policies relating to asbestos, radon, and lead. She argues that a number of early studies and government programs have exaggerated the threats posed by these substances. This book can be ordered at *http://www.cato.org*

Radon

Radon is a naturally-occurring gas that is a source of ionizing radiation. Uranium-238, which is widely distributed in the Earth's crust, decays to radium, which decays to radon. The decay products of radon, known as radon daughters, may attach to dust particles, and are potentially a source of lung cancer, especially among those already at risk, such as smokers. Radon levels vary widely according to geological conditions. High-levels of radon gas have been found in parts of certain states in the Northeast and in the Rocky Mountains.

Concerns about radon were raised based on epidemiological studies of uranium miners who had high exposures to radon, and there is considerable disagreement about the risk associated with lower levels of exposure. In well-ventilated spaces, radon gas is quickly dispersed, but tightly-sealed, energy-efficient housing construction can concentrate the gas in basements and lower floors unless mitigation measures are taken.

Bettis, Clifford, and Carl Throckmorton. "What Teachers Should Know about Radon." ***The Physics Teacher*****, September 1991, pp. 338-343.**
The authors trace significant findings and misconceptions about radon throughout history. They explain the geological conditions associated with radon and potential health effects.

U.S. Geological Survey: Radon in Earth, Air, and Water
This federal government site provides information and maps on regional radon levels in the United States, including an explanation of the geological features enabling radon from underground radioactive materials to pass through the Earth's crust and into the air. ***http://sedwww.cr.usgs.gov:8080/radon/radonhome.html***

Why Files: Radiation Reassessed
Radon was originally advertised as a health cure. This NSF-funded site discusses how perceptions of radiation changed over the years and also includes a clear explanation of uncertainties concerning the "linear, no threshold" model of assessing radiation risk.
http://www.cport.k12.ky.us/FFMS/Curriculum/Science/ScientificInquiry/WHYFILES/how_dang.htm

"EPA Analysis of Radon in Water is Hard to Swallow." ***Science*** **261 (September 17, 1993): 1514-1516.**
This article reviews the debate in Congress and the scientific community over the health hazards of radon in drinking water. It examines the difficulties of merging science and politics in environmental decisionmaking.

Idaho State University: Radiation Information Network
This university site contains articles and other documents on radiation and radon gas, including a comprehensive review of radon science entitled "Radon Update."
http://www.physics.isu.edu/radinf/index.html

Cole, Leonard A. ***Element of Risk: The Politics of Radon*****. New York: Oxford University Press, 1994.**
Cole offers a clear and sometimes amusing analysis of the forces that work in incorporating risk analysis in public policy. He critically examines the role of the press in publicizing the radon scare and describes how radon became a major issue in various U.S. states.

ASBESTOS

Asbestos refers to a group of naturally-occurring minerals which have had wide commercial use because of their resistance to heat and chemical deterioration. Studies of workers with high exposure to asbestos, however, demonstrated that inhalation of fibers was associated with respiratory problems and cancer. Concern about exposure to asbestos used for insulation in schools led to a costly program of removal in many communities, even though the Environmental Protection Agency and other experts advised that the renovation process could produce more exposure to asbestos fibers than would occur if unexposed asbestos insulation were left in place.

Agency for Toxic Substances and Disease Registry. "Public Health Statement: Asbestos." December 1990.
This statement by a division of the Centers for Disease Control explains what asbestos is, how it enters the body, and the health risks thought to be associated with exposures at different levels. Available at *http://www.atsdr.cdc.gov/ToxProfiles/phs9004.html*

Asbestos Hazard Emergency Response Act of 1986
This is the federal legislation that directed the Environmental Protection Agency to establish regulations requiring local education agencies to inspect for asbestos-containing materials in schools. *http://www4.law.cornell.edu/uscode/15/2641.html*

Lee, Martin. "Environmental Protection and the Unfunded Mandates Debate." Congressional Research Service, No. 94-739 ENR, September 22, 1994.
This paper discusses the controversy surrounding asbestos removal in schools. Removal of asbestos from schools across the nation is costly and school districts have incurred large expenditures for removal procedures. Available at
http://www.cnie.org/nle/leg-4.html

Legislation and Policy

Under the provisions of the 1970 Clean Air Act and its subsequent amendments, the U. S. Environmental Protection Agency sets air quality standards for pollutants that reduce their presence to levels that pose no harm to health and welfare. The EPA has established six pollutants that it has designated as "criteria" pollutants: carbon monoxide (CO), lead (Pb), sulfur dioxide (SO_2), ozone (O_3), particulate matter, and nitrogen dioxide (NO_2).

The Clean Air Act was amended in 1990 to include provisions for market-based emissions abatement policies. In the past, pollution control policy (described as a "command and control" approach) prescribed both emissions limits and the technology required to meet those limits. However, since it is costly to build a new plant with the required technology, many utilities continued to use older, more polluting plants. Moreover, regulations that require specific types of emissions control tend to "freeze in place" a technology that may not be the most effective. Market-based approaches, often relying on allowance trading mechanisms, overcome the inefficiencies of command-and-control approaches because they provide incentives to reduce pollutants and the flexibility to innovate and find the most cost-effective methods of controlling emissions.

The Clean Air Act (42 USC Sec. 7401)
The full text of the current federal law governing air pollution appears at this university site. *http://www4.law.cornell.edu/uscode/42/ch85.html*

Environmental Protection Agency: Office of Air and Radiation
This federal agency site provides information on the ozone layer, radiation, urban air, air toxics, acid rain, and vehicle emissions. *http://www.epa.gov/oar.* The EPA's report to Congress, *National Air Quality and Emission Trends Report 1997*, is available at
http://www.epa.gov/oar/aqtrnd97/

The Plain English Guide to the Clean Air Act
This U.S. Environmental Protection Agency guide discusses the main features of the 1990 Clean Air Act and its 1997 amendments.
http://www.epa.gov/oar/oaqps/peg_caa/pegcaain.html

Krupnick, Alan J., and Paul R. Portney. "Controlling Urban Air Pollution: A Benefit-Cost Assessment." ***Science*** **252 (April 26, 1991): 522-527.**
This article is useful as a detailed demonstration of a benefit-cost assessment of a particular environmental policy. The authors estimate the benefits and costs expected to result from the National Ambient Air Quality Standards set under the 1990 Clean Air Act Amendments, both for the nation as a whole and for the Los Angeles area. The authors review the various factors and the uncertainties to be considered.

Bryner, Gary. ***Blue Skies, Green Politics: The Clean Air Act of 1990 and its Implementation.*** **Washington, DC: Congressional Quarterly Books, 1995.**
Bryner provides an account of the political maneuvering that yielded the 1990 Clean Air Act Amendments and a detailed description of the policies that resulted.

Revisions to National Ambient Air Quality Standards

The most controversial regulations to emerge from the 1990 Clean Air Act Amendments have been the 1997 revised national ambient air quality standards (NAAQS) for ozone and small particles. There is disagreement concerning whether the pollutants targeted by the NAAQS present a major threat to public health threat. Ozone, for example, is a temporary problem, subject to certain weather and geographic conditions. The biological mechanism by which certain particles may be harmful is not well understood; it is not known whether it is their size, their composition, or some other factor that might be implicated. The U.S. Court of Appeals for the District of Columbia recently struck down the EPA's regulations on NAAQS, which will reopen the debate on their implementation.

Lave, Lester B. "Environmental Protection Agency's Proposed Air Quality Standards." ***The Brookings Review*** **15 (Summer 1997): 40-47.**
This online publication hosts a pair of articles about the Environmental Protection Agency's air quality standards – one article defends the standards as sensible and the other argues that they represent "the costly pursuit of the impossible." Available at
http://www.brook.edu/PUB/REVIEW/SUMMER97/epa.HTM

National Council for Science and the Environment: Air Issues
This site provides access to Congressional Research Service articles and issue briefings on current clean air legislation and regulation. *http://www.cnie.org/nle/crsair.html*

Chilton, Kenneth, and Christopher Boerner. "Health and Smog: No Cause for Alarm." Regulation 18 (Summer 1995): 50-62.
The authors argue that the federal standard for ground-level ozone, which is based on the fourth highest daily ozone level in a three-year period, tends to overstate general ozone levels. *http://www.cato.org/pubs/regulation/reg18n3f.html*

Emerging Issues

With the Clean Air Act, Congress chose to target manufacturers of automobiles rather than individuals' driving behavior, and to control emissions *rates* rather than actual emissions. Automobile manufacturers were required to meet limits on emissions, with stricter limits on personal vehicles than light trucks and other business vehicles. New cars emit significantly fewer volatile organic compounds and other emissions than 1970 models. In addition, more than half of the sales of new vehicles are of light trucks and vans, which have not been subject to the same strict emissions limits as passenger cars. The Environmental Protection Agency has promulgated new restrictions on sports utility vehicles and vans.

Beaton, Stuart P., et al. "On-Road Vehicle Emissions: Regulations, Costs, and Benefits." *Science* 268 (May 19, 1995): 991-993.
The authors report the results of a study in which remote sensors were used to measure the actual emissions of automobiles on the road in various urban locations in California. They found that fewer than 10 percent of all the cars on the road accounted for 50 percent of the carbon monoxide and hydrocarbon emissions, and that poorly maintained vehicles, new or old, are the most significant sources of these emissions.

Peters, Eric. "They May Be after Your Old Car." *Consumers' Research Magazine*, February 1997, pp. 15-17.
Peters presents the debate over so-called "clunker laws" intended to reduce the number of old (and higher polluting) cars on the road.

Burke, Maria. "Are Oxyfuels Good for Us?" *New Scientist* 147 (July 15, 1995): 24-27.
The author examines the controversy over the use of oxygenated fuels. Oxyfuels contain ethers and alcohols that prevent the build-up of carbon monoxide from automobile exhaust. However, there is some question of whether these additives contribute to other types of pollution.

U.S. Environmental Protection Agency: Cleaner Vehicles and Cleaner Gasoline
This Environmental Protection Agency page includes information on revised air quality regulations that set emission standards for pick-up trucks, minivans, and sport utility vehicles (SUVs) at the same level set for passenger cars, and reduce the sulfur content in gasoline. The final regulation is also included here. *http://www.epa.gov/oms/tr2home.htm*

Labs and Activities

Harte, John. *Consider a Spherical Cow: A Course in Environmental Problem Solving.* Sausalito, CA: University Science Books, 1988.
This book provides a useful introduction to a quantitative approach to analyzing air pollution. The author demonstrates "back of the envelope" calculations of problems related to acid rain, carbon dioxide, and radon, among other environmental problems.

Driscoll, Jerry A. "Acid Rain Demonstration: The Formation of Nitrogen Oxides as a By-Product of High-Temperature Flames in Connection with Internal Combustion Engines." ***Journal of Chemical Education*** **74 (December 1997): 1424-1425.**
Driscoll describes a classroom exercise designed to illustrate the formation of nitrogen oxides, which contribute to acid rain. Available at
http://jchemed.chem.wisc.edu/Journal/issues/1997/Dec/index.html

CHAPTER 5

Global Climate Change

Background

The Earth's climate is naturally variable, fluctuating between warm and cold phases over time scales from centuries to thousands of years. The Earth is currently in a warming period from the Little Ice Age, a cold period in Europe and North America that began in the seventeenth century. Concerns have arisen in recent years that human activities may produce additional climate change by affecting the natural processes that regulate temperature on Earth. In particular, burning fossil fuels and clearing forests have released stores of carbon dioxide into the atmosphere at a faster rate than would occur from natural cycling processes. A major question is the effect that increased concentration of carbon dioxide and other greenhouse gases in the atmosphere might have on climate.

What are the issues?

The Earth's temperature is maintained within a range hospitable to life as the result of a naturally occurring phenomenon termed the "greenhouse effect." This is a warming of the Earth's surface that occurs because some gases in the atmosphere block the transmission of infrared radiation. The Earth absorbs some of the energy carried by the radiation it constantly receives from the Sun and re-emits the remainder as infrared heat radiation. So-called "greenhouse gases" in the atmosphere absorb some of this infrared radiation and reradiate it back to Earth. The net result of this process is a redistribution of energy that makes the surface and lower atmosphere of the Earth warmer, and the upper atmosphere cooler, than they would be otherwise. This warming of the Earth's surface makes our planet much more hospitable to human and other life than it would be if the atmosphere did not contain greenhouse gases.

The most important greenhouse gas is water vapor; other naturally occurring greenhouse gases include carbon dioxide, and methane. Nitrous oxides, and chlorofluorocarbons (CFCs), which have been used as refrigerants and fire retardants, are other greenhouse gases that are released by human activity. All the greenhouse gases together constitute less than 0.1 percent of the atmosphere. Measurements indicate that the CO_2 content of the atmosphere has increased from 280 to 360 parts per million since the beginning of the Industrial Revolution. Increases have also been observed in the concentrations of methane, which is emitted as a by-product of agricultural practices such as livestock rearing and rice farming, and from other sources.

Computer models that simulate the Earth's climate have been used to estimate how much increased concentrations of greenhouse gases might raise the Earth's average temperatures, and what the consequences might be. Perhaps the most important negative effect of higher global temperatures is higher sea levels resulting from thermal expansion of the oceans and possible melting of the Antarctic ice sheet. Other concerns include possible disruption or shifting of rainfall and other regional weather patterns, which would put stress on ecosystems, both managed (such as agricultural cropland) and unmanaged.

What are the uncertainties?

Global warming predictions come from computer simulations that combine theoretical principles and knowledge of past climate patterns. The increase in greenhouse gases from human sources has been measured and is known with some certainty, but less is currently known about other important parts of the subject. Vigorous scientific debates are continuing about how much increased amounts of greenhouses gases from human activities have contributed to a global warming trend. Research has improved understanding of many interactions among oceans, land, biota, and atmosphere that affect climate, but there remain areas of uncertainty, including whether an increase in atmospheric water vapor and clouds would tend to increase or decrease climate warming. Further, current climate simulation models cannot predict the regional effects of a warmer average global climate. For example, while some regions might benefit from warmer, wetter climates, and longer growing seasons, other areas might suffer droughts—the models cannot identify which effects would occur where.

What are the risks and tradeoffs?

The economic growth that has brought prosperity, health, and technological advancement to the United States and other industrialized nations has been driven largely by energy from carbon-intensive fuels. These nations have improved the efficiency of energy use, but fossil fuels remain the major source of energy for transportation and generation of electricity, so that reductions in their use would have substantial impacts. Meanwhile, nations such as India and China, which have vast populations and are trying to rapidly develop their economies, also expect to rely primarily on fossil fuels, which are the cheapest, most readily available energy source. Achieving international agreement to reduce emissions of carbon dioxide and other greenhouse gases has proven difficult.

BASIC CONCEPTS

The body of scientific knowledge about climate change is constantly growing and changing. Many textbooks provide only brief discussions of this complicated topic. Therefore, teachers need to supplement their reading on this topic with other resources. Earth science textbooks can help in this regard, as they often contain one or more chapters on climate and weather. In addition, a number of topical, readable, and highly informative books on climate change have been written in recent years. Many of the books below may be available in public libraries or university libraries. There are also a number of resources available online that will be helpful.

Somerville, Richard. *The Forgiving Air: Understanding Environmental Change*. Berkeley, CA: University of California Press, 1996, Ch. 4, pp. 55-114.
Somerville offers a basic explanation of science related to climate change. He tells the story of early attempts to create mathematical models of the weather and how those evolved into contemporary computer-based models for predicting both short-term weather changes and longer-term climate changes. He also includes a discussion of important climate "feedback effects."

Philander, S. George. ***Is the Temperature Rising? The Uncertain Science of Global Warming.*** **Princeton: Princeton University Press, 1998.**
Philander, a geoscientist at Princeton University, provides a detailed but clear overview of the scientific concepts related to climate change, including the greenhouse effect and the role of winds, weather, and atmosphere. This book is based on Philander's introductory course on climate change and is intended for a lay audience; quantitative details, including equations for calculating the Earth's radiation budget, are reserved for the appendices.

Anthes, Richard. ***Meteorology.*** **Upper Saddle River, NJ: Prentice Hall, 1997.**
This is a concise but detailed textbook that provides a basic introduction to atmospheric sciences, the global energy budget, the greenhouse effect, and other basic concepts.

United Nations Environment Programme and World Meteorological Organization: Common Questions about Climate Change
These two international organizations provide basic answers to the questions commonly asked about climate change, including the evidence upon which scientists are relying for their research in this area, what is not known, and what possible consequences may be. *http://www.gcrio.org/ipcc/qa/cover.html*

Firor, John. W., ed. ***Global Warming: Selected Reprints.*** **College Park, MD: American Association of Physics Teachers, 1995.**
This booklet contains sixteen papers originally published in scientific journals on issues related to climate change. Articles discuss how climate changes may be measured, sea level variability, ice age cycles, and hurricane intensity, among other topics. See *http://www.aapt.org* for ordering information.

Houghton, J. ***Global Warming: The Complete Briefing.*** **Oxford: Cambridge University Press, 1997.**
Houghton, who was co-chairman of the Scientific Assessment Working Group of the Intergovernmental Panel on Climate Change (IPCC), provides a detailed, clear, explanation of the state of scientific knowledge on climate change.

Carbon Cycle and Atmospheric Basics

The Earth system is composed of four great interacting subsystems: the atmosphere (air), hydrosphere (water), lithosphere (land), and biosphere (living things). Energy and matter cycle through these subsystems on both short and long term time scales. Materials necessary for life, such as carbon, nitrogen, sulfur, and phosphorus, cycle through the atmosphere, oceans, land, and biota through such natural processes as photosynthesis, respiration, and decomposition.

Carbon is the essential building block of life on Earth, and carbon dioxide in the atmosphere plays a critical role in regulating the Earth's surface temperature. Carbon in the atmosphere is taken up during photosynthesis by vegetation on land and in the oceans and released through respiration, volcanoes, decaying vegetation, and combustion. Large amounts of carbon are stored in fossil fuels, which are the remains of plants and animals trapped in sedimentary rock. Burning fossil fuels releases additional stores of carbon dioxide into the atmosphere on a much shorter time scale than would occur by natural processes such as weathering and volcanic emissions. Deforestation also releases stores of carbon from vegetation through burning and decomposition. The following resources may be helpful in reviewing basic concepts related to biogeochemical cycles and atmospheric dynamics.

Schmidt, Victor, and William Harbert. *Planet Earth and the New Geoscience*. Third Edition. Dubuque, IA: Kendall/Hunt Publishing Co., 1998, pp. 356-360.
This Earth science textbook includes several detailed chapters on atmospheric dynamics and biogeochemical cycles. This discussion is more quantitative than many other Earth science textbooks; equations for calculating blackbody radiation and heat flow are included.

Thompson, Graham, and Jonathan Turk. *Earth Science and the Environment*. Second Edition. Philadelphia: Harcourt Brace Publishers, 1999, Ch. 15, pp. 347-364.
In this textbook, Thompson and Turk include detailed discussions of the composition of the atmosphere and the physical and chemical processes behind many important atmospheric phenomena.

Jackson, Andrew, and Julie Jackson. *Environmental Science*. Essex, UK: Longman Publishing, 1996, pp. 102-122.
In this textbook, the authors present a detailed discussion of biogeochemical cycling that includes mathematical models of the cycles.

Woods Hole Oceanographic Institution. "Balancing the Budget." *Oceanus*, Spring 1992, pp. 18-28.
For a detailed discussion of carbon sinks and sources, this article from Woods Hole Oceanographic Institution's magazine may be helpful. Available in NOAA's *Global Change Education Resource Guide*. For ordering information, call NOAA's Office of Global Programs at 1-301-427-2089.

The Greenhouse Effect

The greenhouse effect is a natural phenomenon that contributes to a physical environment suitable for life on Earth. The effect was first recognized by mathematician Joseph Fourier in 1824. In the late nineteenth century, Svante Arrhenius theorized that carbon dioxide released in burning fossil fuels was contributing to the greenhouse effect.

The portion of the Sun's radiant energy that is not absorbed, scattered, or reflected (about 47 percent) passes through the Earth's atmosphere and warms its surface. The heated Earth becomes a radiating body releasing long-wavelength radiation into the atmosphere. Some of this radiation is absorbed in the atmosphere by those gases capable of absorbing long-wave radiation and is reradiated downward to warm the Earth's surface.

The atmosphere is a mixture of gases, the most abundant of which are nitrogen (78 percent), oxygen (21 percent), and argon (almost 1 percent). These gases make up almost 99.9 percent of the atmosphere. The atmosphere also includes water vapor, which varies in concentration around the Earth and according to daily weather conditions. There are also trace gases, including carbon dioxide, methane, ozone, nitrogen dioxide, and sulfur dioxide, among others. These gases are rare and make up less than 0.1 percent of the atmosphere, but since, unlike nitrogen and oxygen, they absorb infrared radiation, their concentration in the atmosphere is important. Each greenhouse gas absorbs radiation at different wavelengths, so they vary in their potency.[7]

While CO_2 is the most commonly talked about greenhouse gas, water vapor is the most important in terms of the total amount of infrared energy trapped, so that a small change in the fraction of water vapor in the atmosphere can mask a large percentage change in carbon dioxide. Other important greenhouse gases are methane, nitrous oxide, and chlorofluorocarbons (CFCs). These gases vary greatly in their absorption capacity and their persistence in the atmosphere.

Kastings, James et al. "How Climate Evolved on the Terrestrial Planets." *Scientific American*, February 1988, pp. 90-99.
This article discusses how the carbon cycle on Earth creates a moderate climate, while Venus experienced a runaway greenhouse effect, and Mars has none at all.

Somerville, Richard. *The Forgiving Air: Understanding Environmental Change*. Berkeley, CA: University of California Press, 1996, Ch. 3, pp. 31-54.
Somerville provides a very clear, simple explanation of the greenhouse effect.

Balling, Robert. *The Heated Debate: Greenhouse Predictions vs. Climate Reality.* San Francisco: PPRI Press, 1992, pp. 8-9.
Balling provides the mathematical formulas used to calculate the contribution of various gases to the greenhouse effect.

Environmental Health Center, National Safety Council: Reporting on Climate Change —Understanding the Science
This online guide is intended to explain climate science in clear terms for journalists. It is therefore easy to understand but also includes detailed data on the major greenhouse gases, their sources, and their contribution to the greenhouse effect. Available at *http://www.nsc.org/ehc/guidebks/climtoc.htm*

Pearce, Fred. "Methane: The Hidden Greenhouse Gas." *New Scientist* 122 (May 6, 1989): 37-41.
Concentrations in the atmosphere of methane, a greenhouse gas, have been increasing at a faster rate than concentrations of carbon dioxide in recent years, but scientists are not certain where the extra gas is coming from. Sources of methane are known to include cattle emissions, rice paddies, termites, and decomposing waste but estimates of the contributions from these and other sources are uncertain.

Kunzig, Robert. "Earth on Ice." *Discover*, April 1991, pp. 54-61.
California scientist John Martin originally proposed, half in jest, dumping iron deposits into the Antarctic sea to proliferate the growth of phytoplankton, which would draw carbon dioxide out of the atmosphere and reduce concentrations of this greenhouse gas in the atmosphere. This article discusses why some scientists think the idea is worth trying, while others disagree.

Herzog, Howard, Baldur Eliasson, and Olav Kaarstad. "Capturing Greenhouse Gases." *Scientific American*, February 2000, pp. 72-79.
The authors propose that reductions of carbon dioxide in the atmosphere can be accomplished by sequestering it underground. This has been accomplished successfully since 1996, the authors report, at a North Shore oil and natural gas field off the coast of Norway, where 20,000 tons of carbon dioxide a week are extracted from the natural gas, compressed, and pumped down into a sandstone layer 1000 meters below the seabed.

The greatest challenge lies in developing technologies for separating carbon dioxide from fuel combustion emissions, where it is present in a less concentrated form.

CLIMATE AND WEATHER

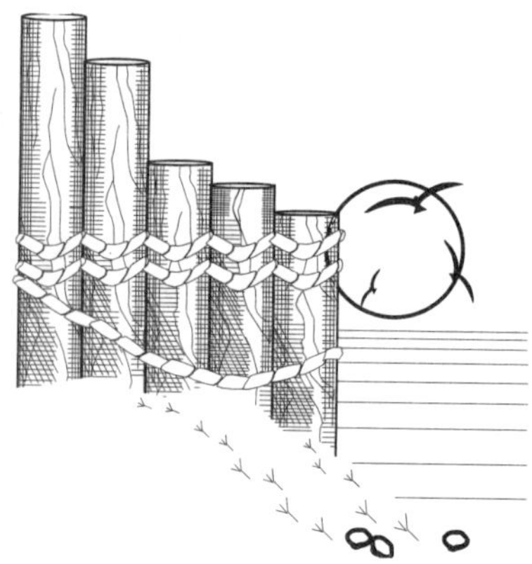

Weather is the short-term pattern of atmospheric conditions, including temperature, precipitation, winds, and humidity, that prevail over time periods of days, weeks, and seasons. Climate refers to average weather considered over a longer term, typically at least thirty years. Global climate is the system that distributes solar energy across the Earth's surface. Uneven heating of the Earth's surface creates wind systems that carry heat from the equator towards the poles. Oceans store solar energy and transport it around the globe on a system of huge deep-water currents known as the "great ocean conveyor belt." The ocean and wind heat-transport systems, and their interactions with each other, are not yet completely understood.

Tarbuck, Edward, and Frederick Lutgens. *Earth Science*. Eighth Edition. Upper Saddle River, NJ: Prentice Hall, 1998.
This textbook includes five chapters on climate and weather, including sections on the composition of the atmosphere, atmospheric moisture, winds, oceans, and climate.

Nese, Jon, and Lee Grenci, eds. *A World of Weather: Fundamentals of Meteorology: A Text/Laboratory Manual*. Dubuque, IA: Kendall/Hunt Publishing Co., 1998.
A World of Weather was written for introductory meteorology courses at the college level. It provides a clear, understandable explanation of basic concepts, and also explains everyday wonders like the "water on the dry road" illusion and the "dog days of summer" folklore.

Webster, Peter, and Judith Curry. "The Oceans and the Weather." In *The Oceans, Scientific American Presents*, Fall 1998, pp. 39-43.
Part of an entire issue devoted to the world's oceans, this article discusses the influence of oceans on weather and climate, including El Niño, hurricanes, monsoons, and the dangerous, quick-developing storms known as "bombs."

Whyte, Ian. *Climate Change and Human Societies*. New York: John Wiley & Sons, Inc., 1996, pp. 37-41.
Whyte's discussion of the Younger-Dryas Event, a rapid cooling that brought a mile-deep layer of ice over Great Britain 12,000 years ago, illustrates the importance of the deep ocean current known as the "great ocean conveyor belt" for the climate of the Northern Hemisphere.

Baskin, Yvonne. "Under the Influence of Clouds." *Discover*, September 1995, pp. 62-69.
This article describes the efforts of the Center for Clouds, Chemistry, and Climate in La Jolla, California, to understand the precise role that clouds play in the Earth's climate system. It discusses theories about the transfer and release of heat energy as water changes form in precipitation and cloud formation.

Monmonier, Mark. ***Air Apparent: How Meteorologists Learned to Map, Predict, and Dramatize Weather.*** **Chicago: University of Chicago Press, 1999.**
This nicely written and illustrated book by a professor of geography at Syracuse University presents a basic course in meteorology by telling the history of how scientists developed weather maps, which, in the days before satellites, was not a simple task.

El Niño and La Niña

The interaction between ocean currents and global weather is demonstrated by the phenomena known as *El Niño* and *La Niña*. Periodically, the westward ocean current off the coast of Peru stalls, producing a buildup of heat energy in the South Pacific waters that affects weather patterns and average temperatures worldwide. This phenomenon is known as *El Niño*. In other years, the westward current is stronger than usual, leading to a dissipation of heat energy in the Pacific. Colder-than-average Pacific Ocean temperatures result, sometimes leading to measurably colder temperatures worldwide. This phenomenon is known as *La Niña*. Together, *El Niño* and *La Niña* make up the Southern Oscillation (or ENSO, for "El Niño/Southern Oscillation"). Recent *El Niño* and *La Niña* weather patterns have complicated efforts to isolate background climate changes that may be occurring.

National Oceanic and Atmospheric Administration: El Niño Page
Devoted to both technical and general-interest information on El Niño and La Niña, this site contains links to educational, forecast, and research sites. The forecasts from the Climate Prediction Center and 3-D graphics animations from the Pacific Marine Environmental Laboratory are impressive. *http://www.elnino.noaa.gov/*

Glantz, Michael. ***Currents of Change: El Niño's Impact on Climate and Society.*** **New York: Cambridge University Press, 1996.**
Glantz describes anomalies in currents and temperatures, human and economic impacts of El Niño-related weather, and the connection between El Niño and climate change.

Fagan, Brian. ***Floods, Famines, and Emperors: El Niño and the Fate of Civilizations.*** **New York: Basic Books, 1999.**
Archeologist Fagan explores how droughts and floods affected ancient civilizations in Egypt, India, Peru, the American Southwest, Central America, and Europe and how unusual climatic events such as El Niño are correlated with major human historical events.

Past Climates: Paleoclimatology

Assessing potential human impact on climate is complicated by the difficulty of isolating human effects from the background of natural climate variability. Indications of past climates are available by examining fossil remains, sediments, and ice cores. The study of historical climate data, called paleoclimatology, attempts to understand the causes and effects of past climate changes to help predict future climate changes.

Whyte, Ian. ***Climate Change and Human Societies.*** **New York: John Wiley & Sons, Inc., 1996, Ch. 1, pp. 9-27.**
Whyte provides basic explanation of methods of studying past climates, including the instrumental record, tree ring measurements, oxygen isotope data taken from ice cores, and the geologic record of the distant past.

Alley, Richard, and Michael Bender. "Greenland Ice Cores: Frozen in Time." ***Scientific American*****, February 1998, pp. 80-85.**
The authors explain how researchers use ice cores to determine the temperature, precipitation, and windiness of past climates.

Beltrami, Hugo, and David Chapman. "Drilling for a Past Climate." ***New Scientist*** **142 (April 23, 1994): 36-40.**
This article discusses how scientists infer past climate changes from the temperature of rock cores at varying depths. The authors also discuss how samples from bore-holes can be used to test surface temperature records measured by thermometers.

National Oceanic and Atmospheric Administration: Paleoclimatology Program
This site includes data and educational information on ancient climates.
http://www.ngdc.noaa.gov/paleo/paleo.html

Illinois State Museum: Ice Ages
This site gives brief discussions of the ice ages, their timing, and their likely causes.
http://www.museum.state.il.us/exhibits/ice_ages/index.html

Hoffman, Paul F., and Daniel P. Schrag. "Snowball Earth." ***Scientific American*****, January 2000, pp. 68-75.**
This article discusses evidence that the Neoproterozoic ice age was so severe that glaciers extended even to the tropics. The authors present a detailed account of the potential climate feedback effects, such as albedo and rain, that researchers believe contributed to replacing a frozen "snowball Earth" with a hot, wet one. This article's clear explanation of climate feedbacks may be helpful in understanding issues related to climate changes.

Predicting Future Climates

Computer climate models (called general circulation models) are mathematical models based on those used to predict local weather patterns. High speed supercomputers are required to process the many variables and equations necessary to simulate future climate changes. The global climate system is complex, and the role of positive and negative feedbacks is not yet completely understood. In particular, the roles of (and interactions between) oceans and clouds are still a matter of uncertainty, and different computer models treat these factors differently with varying results. Global temperatures have not risen as high as climate models have predicted, and this has led scientists to look at other factors, such as sulfate aerosols, which are now believed to produce a cooling effect that may counterbalance greenhouse warming. As more is understood about these phenomena, computer models are refined to take them into account.

Somerville, Richard. ***The Forgiving Air: Understanding Environmental Change*****. Berkeley, CA: University of California Press, 1996.**
Somerville includes a useful history of attempts to model weather and patterns, and a basic explanation of global climate models.

National Aeronautics and Space Administration, Goddard Institute for Space Studies: Global Climate Modeling
This site includes discussions of NASA's own climate model along with papers on other aspects of climate modeling and climate change. Some of these papers are fairly technical. *http://www.giss.nasa.gov/research/modeling/*

Barron, Eric J. "Climate Models: How Reliable are Their Predictions?" *Consequences*, Fall 1995.
Barron, an atmospheric scientist who serves on the National Research Council's Committee on Global Change Research, summarizes in this article the conclusions of a forum, convened at the request of Congress and administration science advisors, to develop "a statement on the credibility of modeled projections of climate change." The article provides a useful overview of the state of climate modeling science as well as an explanation of how the models work. *http://gcrio.ciesin.org/CONSEQUENCES/fall95/mod.html*

Oak Ridge National Laboratory: Carbon Dioxide Information Analysis Center
This center is a government depository for data related to the global carbon cycle and climate change. In the site's FAQ sections, it is possible to retrieve a table indicating data and source of greenhouse concentrations, an illustration of the carbon cycle, an analysis of the world's ecosystems ranked by carbon in live vegetation, plus scientific papers and other data. *http://cdiac.esd.ornl.gov/*

U.S. Global Change Research Information Office
This site, hosted by the U.S. Global Climate Research Program, provides access to data and information on climate change research. This is a clearinghouse for documents and reports on climate generated by participating federal agencies. *http://www.gcrio.org/*

Areas of Uncertainty

It has been established with reasonable certainty that the concentration of carbon dioxide and other greenhouse gases has increased in the atmosphere since the beginning of the Industrial Revolution and that these gases are capable of absorbing infrared radiation. There are a number of natural processes that are not yet well understood. There is, for example, the case of the "missing" carbon. Only about half of the carbon dioxide contributed to the atmosphere from fossil fuel use and land-use change has remained in the atmosphere. It is thought that the other half has been removed by increased photosynthesis from forest regrowth and by the oceans, but the relative contributions of these two sinks are uncertain. The magnitude and timing of any potential warming and its effects on particular geographic areas are also matters of debate.

Barron, Eric. "Global Change Researchers Assess Projections of Climate Change." *Eos*, May 2, 1995, pp. 185, 189-90.
This report, prepared for the General Accounting Office in 1994, identifies areas of climate science about which researchers are virtually certain, explains areas of continuing uncertainty, and suggests opportunities for early progress in reducing uncertainties. Available at *http://www.agu.org/sci_soc/barron.html*

National Research Council. *Reconciling Observations of Global Temperature Change*. Washington, DC: National Academy Press, 2000.
This is the report of a panel convened to examine the differences in observations of temperature at the surface of the Earth, where a warming trend has been measured, and in the upper air, where satellite and radiosonde measurements indicate a lesser warming trend. There are, however, a number of uncertainties that must be resolved to reconcile the two sets of data. Available at *http://www.nap.edu*

Global Hydrology and Climate Center
This center, a joint venture between government agencies and academia, provides explanations of satellite-based temperature-measuring instruments known as Microwave Sounding Units. *http://www.ghcc.msfc.nasa.gov*

Center for International Earth Science Information Network: The Use of Satellite Remote Sensing
This online guide helps educators find selected key documents and data sets for the use of satellite remote sensing to study global environmental change.
http://www.ciesin.org/TG/RS/RS-home.html

Charlson, Robert, and Tom Wigley. "Sulfate Aerosols and Climate Change." ***Scientific American*, February 1994, pp. 48-57.**
The authors discuss evidence that industrial emission of sulfur compounds and other man-made aerosols, which reflect sunlight back into space, explains why Earth has not warmed as dramatically as predicted in computer climate models.

Suplee, Curt. "Untangling the Science of Climate." ***National Geographic*, May 1998, pp. 38-71.**
This feature article provides a basic introduction to climate variability and climate history, computer climate models, the role of different factors such as oceans, and forcing mechanisms such as anthropogenic greenhouse gases and volcanic dust.

Ciais, P. et al. "A Large Northern Hemisphere Terrestrial CO_2 Sink Indicated by the 13C/12C Ratio of Atmospheric CO_2." ***Science* 269 (August 25, 1995): 1098-1102.**
Approximately half of all carbon dioxide emissions in the Northern Hemisphere do not remain in the atmosphere, but there has not been sufficient data available to assess the amount of this carbon dioxide that is absorbed by the oceans and by the terrestrial biosphere. This article (a challenging one) describes research that suggests a significant amount of carbon dioxide is taken up by land in the Northern Hemisphere, perhaps because of forest regeneration that has occurred in the last few decades.

Pearce, Fred. "Greenhouse Wars." ***New Scientist* 155 (July 19, 1997): 38-43.**
The author interviews several prominent scientists who have raised questions about some of the assumptions that are used in climate models to predict future climate and who point to the discrepancy between satellite data, which do not indicate a consistent warming trend, and surface measurements as evidence that current models are flawed.

Climate Change Policies

Policymakers in the United States and other countries are seeking agreement on measures to reduce the greenhouse gases produced by human activity and to mitigate possible effects on climate. The 1997 Kyoto Protocol on climate change, which has been signed by many countries (but has not been ratified to date as a treaty in the United States), seeks to commit the governments of the advanced industrial nations to reduce emissions of carbon dioxide and other greenhouse gases. One serious disagreement is that large, rapidly developing countries such as India and China would not be subject to restrictions, even though their energy use will dwarf that of the United States in coming years. The Kyoto Protocol proposes mechanisms to encourage developed nations to transfer energy and pollution-control technologies to developing countries.

United Nations Framework Convention on Climate Change
The most important worldwide organization devoted to addressing the threat of climate change, this web page includes reports on the most recent meetings, links to treaty texts, and policy proposals. *http://www.unfccc.de*

National Council for Science and the Environment: Climate Change
This site presents a collection of articles for the non-specialist on climate change policy, including topics such as the history of United Nations climate agreements, assessments of different policy approaches, and summaries of key documents.
http://www.cnie.org/nle/crsclim.html

Resources for the Future
Resources for the Future maintains two separate web sites devoted to climate change policy: its Climate Page and a page entitled Weathervane. These sites contain articles and research reports analyzing current policy proposals for mitigating anthropogenic climate change. *http://www.rff.org/environment/climate.htm* and *http://www.weathervane.rff.org*

Policy.com: Issue Archive
An Internet clearinghouse for information and opinion on various public policy issues, *Policy.com* has several features on climate change from a variety of viewpoints.
http://www.policy.com

Moore, Thomas Gale. *Climate of Fear: Why We Shouldn't Worry about Global Warming*. Washington, DC: Cato Institute, 1998.
Economist Moore argues that the harms arising from a warmer climate have been overstated and the benefits understated. He argues that the specific threats from a warmer climate could be addressed by technological countermeasures, such as building dikes to protect threatened low-lying areas, whereas stringent limits on energy use would impose significant hardships. This book can be ordered at *http://www.cato.org*

LABS AND ACTIVITIES

***Global Change Education Resource Guide*. Office of Global Programs, National Oceanic and Atmospheric Administration, 1998.**
This guide, available free from the National Atmospheric and Oceanic Administration, includes fact sheets, overheads, and articles from scientific journals on natural climate variability and past climates. Available by calling 1-301-427-2089.

NASA Discover Earth: Earth's Energy Budget or Can You Spare a Sun?
This project provides a lesson plan that helps students visualize and calculate the Earth's energy budget – the equilibrium between the radiation received by and the radiation emitted by the Earth. *http://www.strategies.org/Lesson8.html*

NASA Discover Earth: Albedo Versus Temperature
This project has a lab activity using satellite data to examine the relationship between cloud cover and albedo (the fraction of solar energy that is reflected back to space) and local temperature. *http://www.strategies.org/lesson3.html*

Bohren, Craig. *Clouds in a Glass of Beer: Simple Experiments in Atmospheric Physics.* New York: John Wiley & Sons, Inc., 1987.
Bohren offers several experiments that are designed to be entertaining and faithful to the rigorous science of atmospheric physics.

Background

Throughout history humans have discovered a variety of ways to convert nature's energy sources to use. Fire was first used for heat, cooking, and clearing land. Animals were domesticated some 5000 years ago to supplement human muscle power for plowing and transportation. Wind power has been used to mill grain and propel ships for thousands of years. It was not until the Industrial Revolution, though, that large-scale use of fossil-fueled machinery began to replace human labor for agriculture and manufacturing. The development of coal-powered steam engines in the mid-1700s made it possible to manufacture the high-quality tools needed to build machines and to mechanize agriculture. The efficiencies permitted by mechanized farming and manufacturing enabled the transformation of Western countries from largely agrarian societies to urbanized populations with sufficient leisure and wealth to increase literacy rates and to spur the scientific and technological advances that are the foundation of advanced civilization.

What are the issues?

Energy is critical to the industrialized economy. It is necessary to process, manufacture, and transport goods; to heat, cool, and light homes; and to power communication equipment. This energy is currently provided predominately by the use of fossil fuels such as oil, coal, and natural gas. Fossil fuels account for about 90 percent of world energy consumption.[8] Petroleum leads with a share of about 40 percent of total world energy consumption, followed by coal (24 percent) and natural gas (22 percent).

However, extraction, transport, and combustion of fossil fuels exact a toll on the environment. Combustion of fossil fuels produces carbon dioxide, a greenhouse gas that has been implicated as contributing to climate change. Incomplete combustion of fossil fuels also results in noxious emissions of carbon monoxide and other air pollutants. Fossil fuels are not easily replaced. Unlike renewable sources such as wind energy and solar energy, fossil fuels are easily stored and transported within the energy systems that have evolved around them. Further research is required to develop energy technologies that satisfy the growing demands for clean energy and also replicate the versatility that has made fossil fuels the basis of contemporary civilization in most of the world.

What are the uncertainties?

Past predictions that oil reserves would disappear by the year 2000 have been revised as new reservoirs are discovered and enhanced extraction technologies are developed. Since the oil shocks of the early 1970s, the price of oil dropped. This has hindered developments in renewable energy technologies such as wind and solar, which have been unable to compete with abundant, inexpensive, and easily stored and transported petroleum. Even with advanced exploration and recovery techniques, crude oil production may begin to decline in the 21st century. This, however, will likely spur advances in recovery methods for unconventional petroleum sources such as oil shale and tar sands. Coal reserves are abundant, but environmental costs arising from mining and burning coal raise concerns about reliance on this energy source.

Energy and Power

1 kilowatt-hour = 3,413 Btus = 860,421 calories
1 Btu = 0.000293 kilowatt-hour = 252 calories = 1.055 joule
1 watt = 3.413 Btu/hr = 14.34 calories/min
1 calorie = the amount of heat necessary to raise the temperature of 1 gram (1 cm^3) of water 1 degree Celsius
1 quadrillion Btu = (approximately) 1 exa joule
1 joule = 0.239 calorie = 2.778×10^{-7} kilowatt-hour

What's a joule? Science education Michael Chapin suggests that the joule is an extremely intuitive unit: "To get a feel for it, pick up a liter bottle of water and toss it ten centimeters into the air. You've just expended about a joule." [The Universe in the Classroom 43 (Third Quarter 1998), p.3]. For easy conversion of measurements:
http://www.megaConverter.com/

The trend among industrialized countries has been toward increasing energy *intensity*, that is, using energy more efficiently in manufacturing, heating, cooling, and other processes. Gains have also been made in controlling emissions. Developing countries, however, are expanding their consumption of fossil fuels, but without the emissions control technologies and advances in energy conservation that industrialized nations have achieved. There is disagreement concerning whether rapidly industrializing countries can "skip over" the long transition to clean technologies that occurred in Western nations and mitigate the environmental impacts of development.

What are the risks and tradeoffs?

Energy use and income levels are, in most cases, tightly linked and vast amounts of energy, especially electricity, are required to support a high qualify of life in modern industrialized nations. We have become so reliant on the benefits of a plentiful, cheap energy supply that it is hard to imagine the lack of it. It is difficult to envision, for example, what a hospital would be like without electricity or any of the technologies that electricity has made possible. For many people in undeveloped countries, such is their daily life.

Developing countries are increasing their industrial capacity and use of energy in order to improve the quality of life of their populations. It is a major challenge to supply the energy necessary to support a decent lifestyle for a growing global population in ways that minimize environmental damage.

Nuclear technology was once believed to be the solution to finding a reliable energy source, but fears about the safety of nuclear power plants and the disposal of nuclear wastes have halted production of reactors in this country. Research continues in renewable energy technologies such as solar cells and wind turbines. However, these renewable energy technologies do not yet offer the power outputs required to contribute significantly to the overall energy supply, and the inability to store and transport energy from these sources poses additional challenges. There are environmental costs associated with alternative sources as well. Hydroelectric power, for example, imposes significant impacts on river ecosystems. Personal transportation and transportation of food and agricultural products are critical elements of the industrialized world, but manufacturers have not yet developed an alternative to the fossil-fuel powered engine that is widely available at an affordable cost.

Basic Concepts

"In this house, we obey the laws of thermodynamics!"

★Homer Simpson★

Energy is defined as the ability to do work. All biological systems are sustained by the conversion of radiant energy from the Sun into usable chemical energy. Introductory physics textbooks provide explanations of basic concepts of energy flows and the laws of thermodynamics. In addition, there are several books that discuss energy and energy systems in clear, simple terms. Many of the following books and articles can be found in a public library or through online databases.

Christensen, John. *Global Science: Energy, Resources, Environment.* Dubuque, IA: Kendall/Hunt Publishing Co., 2000.
Christensen discusses energy and energy systems, energy efficiency, the laws of thermodynamics, and energy flow in ecosystems. The textbook also includes discussions of energy history, the formation and makeup of fossil fuels, and important aspects of the discovery and recovery of energy resources.

Goldstein, Martin, and Inge Goldstein. *The Refrigerator and the Universe: Understanding the Laws of Energy.* Cambridge, MA: Harvard University Press, 1995.
The authors explain the laws of energy in both technical and everyday terms. They devote three chapters to the second law of thermodynamics.

Cassedy, Edward, and Peter Grossman. *Introduction to Energy: Resources, Technology, and Society.* New York: Cambridge University Press, 1990.
The authors use diagrams to describe, in basic terms, the workings of conventional energy conversion systems – steam engines, hydroelectric turbines, internal combustion engines, and gas turbines. They also provide an appendix on specific scientific concepts and principles that pertain to energy: units of measure, forms of energy, laws of thermodynamics, and laws governing heat engines.

Smil, Vaclav. *Energy in Human History.* Boulder, CO: Westview Press, 1994.
Smil begins this book with a brief survey of energy terms and concepts, and provides appendices on energy density, intensity, and efficiency. He provides a detailed account of the history of human energy use before the Industrial Revolution, including discussions of prehistoric energy use, traditional agriculture, and pre-industrial energy technologies and fuels.

Thorndike, Edward. *Energy and Environment: A Primer for Scientists and Engineers.* Reading, MA: Addison Wesley Publishing Co., 1976.
This basic introduction to energy issues is for an advanced audience. The book is currently out of print, but may be found in university libraries.

Von Baeyer, Hans Christian. ***Maxwell's Demon: Why Warmth Disperses and Time Passes*****. New York: Random House, 1998.**
In this interesting and accessible book, physics professor Von Baeyer tells the history of the discovery of the second law of thermodynamics and how simple questions such as "What is warmth?" lead to scientific discoveries. He explains heat, energy, and entropy and addresses questions such as why gasoline-fueled engines, at their best, can run at only 30 percent efficiency.

International Energy Agency
This site, from an international energy forum, gives both statistics and clear charts on energy consumption and production patterns, by country and region, source, and economic sector. *http://www.iea.org*

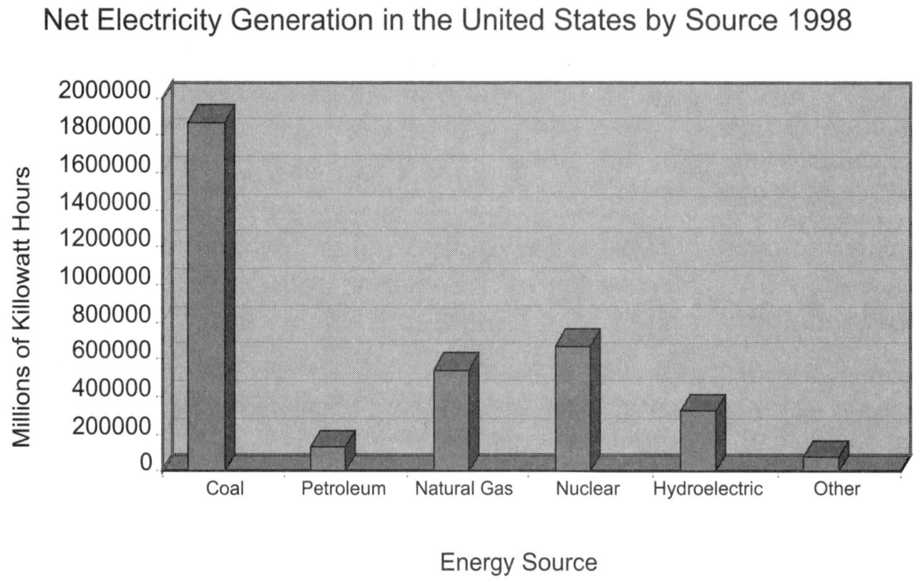

Source: U.S. Department of Energy, *Electric Power Annual 1998.*

ELECTRICITY

Electric power has dramatically transformed human life and society. The development of electric motors revolutionized industry in the early 1900s, bringing about rapid advances in productivity. More recently, electricity has powered the information revolution. Electricity for home appliances has enormously reduced the amount of time required for cooking, cleaning, and other life-sustaining chores. For the one-third of the global population that does not have access to electrical power, this lack is significant. In undeveloped countries, time spent for survival activities can be as much as nine to ten hours a day. This burden falls most heavily on women and children, who, because their labor is required for survival, are less likely to receive an education. Moreover, reliance on wood and other biomass materials for heating and cooking contributes to deforestation in many areas. Lack of access to electricity is a significant barrier to easing poverty and environmental degradation for undeveloped countries (see Population chapter for more on the "vicious circle" of poverty and environment).

Most electricity (53 percent) in the United States is generated by coal power plants, according to the Department of Energy's *Electric Power Industry: 1998 Review.* Eighteen percent is generated by nuclear power plants and 14 percent is supplied by natural gas generating plants. Hydroelectric dams supply approximately 10 percent. Renewable energy sources other than hydroelectric, such as solar, wind, and biomass, generate only about 2 percent of U.S. electrical supply.

Electricity is distributed throughout the United States by an electric power system that connects generation, transmission, distribution, and other facilities. The flow of electricity through the system is maintained by dispatchers who match the supply of electricity with the demand for it, so that electricity is available whenever a customer flips a switch.

Energy Information Agency: Electric Power Industry Overview
This U.S. Department of Energy office provides a brief overview of how electricity is generated and transmitted in the United States. The site also explains the interconnected power grids that transfer electricity from one utility company to another as demand requires. Also available on the EIA site are statistical reviews of the current outlook in electricity generation by source, country and region, as well as projections of future use patterns extending to 2015.
http://www.eia.doe.gov/cneaf/electricity/page/prim2/toc2.html

Sustainable Energy and Environment Division. "Energy After Rio: Prospects and Challenges." United Nations Development Programme, 1997.
Chapter Two of this extensive report provides a detailed assessment of the link between energy use and per capita income, including the amount of time spent on survival activities for several developing countries, and the correlation between access to electricity and literacy. Available at *http://www.undp.org/seed/eap/rio/contents.html*

Foran, Jack. "The Day They Turned the Falls On: The Invention of the Universal Electrical Power System." ***Case Studies in Science.***
In this interesting historical case study, the author discusses how the universal electrical power system resulted from a struggle over the proper way to harness the power of Niagara Falls. Available at *http://ublib.buffalo.edu/libraries/projects/cases/niagra.htm*

Easterbrook, Gregg. "The Future of Electric Power." ***The Atlantic Monthly,*** **July 1993, pp. 42-46.**
According to the author, deregulation in electric utilities may lead to smaller-scale, more efficient ways of producing and consuming electricity.

Energy Conservation

Using energy more efficiently reduces not only emissions, but costs. There is therefore a significant incentive to improve energy efficiency in manufacturing processes, construction of new buildings, and appliances. The trend in the U.S. over the past few decades is increasing energy *intensity*, that is, less energy used per unit of Gross Domestic Product (GDP). Energy conservation is not entirely without environmental impacts, however. Some new energy-efficient technologies, such as superconductors, rely on toxic materials. Energy efficient lighting uses mercury vapor; and light-emitting diodes (LEDs), which might someday replace the light bulb, use arsenic or gallium. These are factors to consider in assessing the comparative environmental impact of these technologies and demonstrate one of the key concepts in environmental science: There's no such thing as a free lunch. There are always tradeoffs to consider.

Lawrence Berkeley National Laboratories: Center for Building Science
This government research site discusses new energy efficient technologies for windows and lighting, tools for calculating energy inefficiency in buildings, and links to other web resources on energy efficiency. *http://eetd.lbl.gov*

Fickett, A., C. Gellings, and A. Lovins. "Efficient Use of Electricity." ***Scientific American*****, September 1990, pp. 65-74.**
The authors discuss strategies for conserving energy and decreasing the demand for electricity rather than increasing the supply.

American Council for an Energy Efficient Economy
This organization concludes that energy efficiency is good, not only for the environment, but also for business profits and national security. Their site includes articles on the benefits of improving energy efficiency, in addition to statistics and graphs illustrating energy use patterns in the United States. *http://www.aceee.org*

Hirst, Eric. "A Bright Future: Energy Efficiency Programs at Electric Utilities." ***Environment*****, November 1994, p. 10**
The author details the "demand-side management" (DSM) programs that many utilities rely on to encourage their customers to use energy more efficiently, such as providing rebates for purchasing more efficient building materials and appliances. The bad news, according to the author, is that DSM programs have not been as cost-effective as predicted; they can, however, save a utility the cost of building new power plants to meet increasing demand.

McConnell, Robert L., and Daniel C. Abel. "Cold Facts about Energy Waste and Avoidable Air Pollution." In ***Environmental Issues: Measuring, Analyzing, and Monitoring*****. Upper Saddle River, NJ: Prentice-Hall, 1999, pp. 116-125.**
This exercise asks students to calculate the cumulative effect of seemingly insignificant uses of energy, such as the amount of electricity generated to make ice or power the clock on VCRs (that are blinking in many homes because, according to the U.S. Census bureau, 20 percent of VCR owners report that they do not know how to program the clock).

Fossil Fuels

The three main fossil fuels are petroleum, coal, and natural gas. Fossil fuels were formed after absorption of radiant energy from the Sun by plants long ago. This radiant energy promoted plant growth as the result of well-known photochemical processes. This energy was stored either in the remains of plants themselves, or in the bodies of animals that had consumed them. These remains were attacked by anaerobic bacteria that consumed their oxygen, leaving only molecules of hydrogen and carbon (hydrocarbons). These molecules were compressed and heated over time by sedimentary layers that built up over the deposits.

Reservoirs of fossil fuels are designated as resources or reserves. Resources are those supplies that are known to exist. Reserves are those resources thought to be recoverable economically using existing methods. While almost 57 percent of known petroleum reserves are thought to be located in the Middle East, coal reserves are more evenly distributed among North America, Europe, and Asia. The largest natural gas reserves are located in the Middle East and the former Soviet Union.

Near-term concerns about fossil fuel supplies focus on petroleum, production of which is predicted to decline beginning some time in the 21st century. Alternative forms of petroleum such as tar sands and oil shale may extend supplies as technologies are developed to exploit these resources economically. Reserves of natural gas and coal are expected to last longer at current rates of use.

"Special Report." ***Scientific American*****, March 1998.**
This issue of Scientific American contains a collection of articles on fossil fuels. These articles cover the uncertainty of future oil production, the prospects for mining unconventional sources such as tar sands, advances in oil drilling technology, and the prospects for natural gas-based liquid fuels.

Schmidt, Victor, and William Harbert. ***Planet Earth and the New Geoscience*****. Third Edition. Dubuque, IA: Kendall/Hunt Publishing Co., 1998.**
In Unit 12 of this book, the authors describe the formation and geologic features of fossil fuel deposits, their worldwide distribution, and the technologies for extracting them.

U.S. Geological Survey, National Oil and Gas Resource Assessment Team. ***1995 National Assessment of United States Oil and Gas Resources*****. U.S. Geological Survey Circular 1118. Washington, DC: U.S. Geological Survey, 1995.**
This booklet summarizes the findings of a three-year assessment of oil and gas resources in the United States and presents estimates of actual reserves, future reserves, and potential resources. This report can be ordered from 1-888-ASK-USGS.

PETROLEUM

The twentieth century was a time of unparalleled growth in economic activity, travel, exploration, and scientific discovery made possible by technologies based on petroleum. More than half of the world's petroleum is used for transportation (automobiles, trucks, and other vehicles). Petroleum is also the primary source for the manufacture of plastics and many critical chemicals, and there is currently no substitute for its use in these processes.

Concerns have been raised about supplies of petroleum, but current estimates indicate that the world has large reserves, despite increasingly large extractions of oil. The largest proven reserves are in the Middle East, which raises some issues related to U.S. dependence on imports for a critical energy source. The more pressing concern is that combustion of petroleum results in emissions of air pollutants and particulate matter, and carbon dioxide, a greenhouse gas.

Oil History
This online book by petroleum geologist Samuel T. Pees focuses on the history of oil exploration beginning in 1859 at Oil Creek in Titusville, Pennsylvania.
http://www.oilhistory.com

U.S. Department of Energy, Office of Fossil Energy: Oil and Gas Programs
This federal government page includes articles on extraction and processing of petroleum. It also discusses the emergency supply of oil, known as the Strategic Petroleum Reserve, that the department maintains in giant salt caverns along the Gulf Coast of Texas and Louisiana. *http://www.fe.doe.gov/programs_oilgas.html*

"Special Report." ***Scientific American*****, March 1998.**
This special issue includes a number of articles on petroleum, including "Oil Production in the 21st Century," which describes the sophisticated technology used to extract oil; "The End of Cheap Oil," which raises questions about means for estimating proven oil

reserves; and "Mining for Oil," which assess the prospects for augmenting petroleum resources using oil found in "tar sands," a geologic formation in which bitumen, a tar-like form of oil, is mixed with sand.

Stanislaw, Joseph, and Daniel Yergin. "Oil: Reopening the Door." ***Foreign Affairs*** **72 (September/October 1993): 81-93.**
Oil will remain a crucial energy source subject to intense competition and innovation, the authors argue. They survey the recent history of petroleum and its future prospects. This journal can be found in a university library or in an electronic periodical database.

Australian Institute of Petroleum: Explore Australia's Petroleum Industry
This industry site offers a basic explanation of how gasoline is refined, including the chemical reactions that occur. *http://www.aip.com.au/frames/education.html*

U.S. Energy Information Administration: Petroleum
This federal government site provides data on U.S. petroleum reserves and articles analyzing aspects of the oil industry.
http://www.eia.doe.gov/oil_gas/petroleum/pet_frame.html

Daniloff, Ruth. "Waiting for the Oil Boom." ***Smithsonian*****, January 1998, pp. 25-35.**
This article describes how development of oil fields near the Caspian Sea is bringing wealth and change to the city of Baku, the capital of Azerbaijan, which has been an "oil town" since ancient times. For related links see
http://www.smithsonianmag.com/smithsonian/issues98/jan98/baku.html

U.S. Environmental Protection Agency, Oil Spill Program: Responding to Spills
This federal government site discusses methods employed to treat water, land, and animals that have been affected by oil spilled into oceans, lakes, and rivers.
http://www.epa.gov/oilspill/response.htm

Romm, Joseph, and Charles Curtis. "Mideast Oil Forever?" ***The Atlantic Monthly,*** **April 1996, pp. 57-74.**
According to the authors, the United States is lowering its preparedness for the next oil crisis: reducing research funding for alternative energy technology just when the technology is showing great promise.

National Council for Science and the Environment: Energy Issues
These Congressional Research Service briefs on various aspects of U.S. energy policy alternatives, include discussions of petroleum supplies and the Strategic Petroleum Reserve. *http://www.cnie.org/nle/crseng.html*

Yergin, Daniel. ***The Prize: The Epic Quest for Oil, Money, and Power.*** **New York: Touchstone Books, 1993 (reissued).**
This is a richly detailed, comprehensive history of the oil industry. The author won the Pulitzer Prize for this account of the impact oil has had on the world.

COAL

Coal is the most abundant fossil energy source and currently the least expensive of the fossil fuels. Coal is now the most widely used fuel for generating electricity in the world, accounting for about 33 percent of total worldwide electricity production. Although coal is widely distributed throughout the globe, over half the recoverable reserves are found in China, the United States, and the former Soviet Union. Coal mining and combustion have become highly regulated, technical operations to mitigate environmental impacts. New "clean-coal" technologies are being developed to offset the environmental impacts of coal combustion.

U.S. Energy Information Administration: Coal
This federal government page contains data and information on coal production, use, and reserves, as well as several articles on the history and current status of coal use in America today. *http://www.eia.doe.gov/fuelcoal.html*

U.S. Department of Energy, Office of Fossil Energy: Coal and Power
This federal government page includes short documents on clean coal technology, such as advanced coal combustion and coal liquefaction methods. Their Clean Coal Technology Projects page also includes reports on electricity plants fired by clean coal technology. *http://www.fe.doe.gov/programs_coalpwr.html*

Kentucky Geological Survey: Coal Information
This university site contains short discussions of the nature, origins, and uses of coal, as well as links to sites with data on coal use, production, and reserves.
http://www.uky.edu/KGS/coal/webcoal/pages/coal3.htm

Ohio State University: Coal Mining in the Gilded Age and Progressive Era
Tracing the history of coal mining in Ohio, this university site includes a number of documents and narrative accounts of coal mining from the late 19th and early 20th centuries, including a 1894 magazine article by Stephen Crane, author of *The Red Badge of Courage*.
http://www.cohums.ohio-state.edu/history/projects/Lessons_US/Gilded_Age/Coal_Mining

Britton, Peter. "Cleaning Up Coal." *Popular Science*, April 1993, pp. 78-83.
The author notes that some are enthusiastic about the prospects for clean coal technologies, while others are convinced that coal will never be clean enough.

Cullen, Robert. "The True Cost of Coal." *The Atlantic Monthly*, December 1993, pp. 38-52.
According to the author, coal is considered a cost-competitive energy option only because major costs, such as those associated with mining and air pollution, are not included in the price of coal.

Copp, Newton, and Andrew Zanella. *Discovery, Innovation and Risk: Case Studies in Science and Technology*. Cambridge, MA: MIT Press, 1993, pp. 129-145.
Copp and Zanella give a brief history and technical overview of the coal-fired steam turbine.

Natural Gas

Natural gas is a mixture of gases, the highest proportion being methane (CH4). As with other fossil fuels, natural gas comes from ancient decomposed plants and animals, and is usually found with deposits of petroleum. For many years, natural gas released during drilling of oil wells was vented off or burned (and continues to be at some sites). Natural gas currently supplies about 21 percent of the world's energy consumption. It is the "cleanest" fossil fuel, emitting less carbon dioxide per unit volume than oil or coal. Natural gas is, however, highly flammable in both liquid and gaseous forms. It can be transported as a gas through pipelines, and over greater distances it can be shipped in liquid form, but heavy tanks are required to keep it at sufficiently high pressure. In addition, leaks from natural gas pipelines add methane, a greenhouse gas, to the atmosphere.

U.S. Department of Energy, Office of Fossil Energy: Oil and Gas Programs
This federal government site includes publications outlining up-to-date recovery techniques for gas reserves. *http://www.fe.doe.gov/programs_oilgas.html*

U.S. Energy Information Administration: Natural Gas
This federal government page supplies data on natural gas production, transportation, and consumption. *http://www.eia.doe.gov/oil_gas/natural_gas/nat_frame.html*

Transportation

Let's face it, there isn't a 16-year-old on the planet who wouldn't eat worms for a red convertible.

★Scott Oldman, "A Century of Cars," **Popular Mechanics Online**★

At this point we digress from the discussion of the major energy sources to explore a major end use of energy. Transportation is perhaps the most challenging end use, because it requires that energy be available in a portable form.

Approximately half of all petroleum refined is used for motor vehicles and their infrastructure. Motor vehicles are an integral part of American life. Trucks, vans, and other vehicles have become crucial for the delivery of goods and the personal automobile has enormously increased personal freedom, convenience, and mobility.

The internal combustion engine was first developed in the late 1880s in Germany. Within a few decades, the discovery of large supplies of oil in Texas and the development of mass production manufacturing methods made automobiles affordable for large numbers of the U.S. population. By 1915, automobiles began to eclipse horses and other draft animals as the predominant mode of transportation in the United States. A vast infrastructure of roads, highways, bridges, gasoline refueling stations, and other services was built to support this new mode of transportation. There are now approximately 600 million motor vehicles worldwide, of which about 75 percent are personal automobiles. The United States leads the world in percentage of car ownership, at about 0.8 cars per capita (including light trucks), but percentages are growing worldwide.[9] It is projected that there will an increase of 30 percent in the worldwide number of motor vehicles within the next decade.

In response to the Clean Air Act of 1970 and concerns about oil prices and availability, automobile manufacturers instituted a number of engineering improvements in order to control emissions and increase fuel economy. These refinements include greater use of lightweight materials, development of emissions control technologies such as catalytic converters, and use of computer-sensing equipment to monitor engine performance. As a result, average miles per gallon for passenger vehicles has increased significantly since the 1970s and a new, well-maintained automobile releases about 95 percent fewer harmful emissions than cars of two decades ago.[10] Nonetheless, certain emissions remain a concern, including nitrous oxides, which contribute to formation of acidic precipitation in areas where traffic is heavy, and carbon dioxide, a greenhouse gas, which is a by-product of the combustion of fossil fuels. Transportation accounts for about 30 percent of all human-produced contributions of greenhouse gas in the United States.[11]

Research continues, therefore, to develop transportation technologies that reduce emissions. One strategy is continuing improvements in engine performance. For example, only about 12 percent of the energy produced in a typical car is converted into usable kinetic energy. Design improvements in this area would increase the number of miles per gallon. Other technologies are discussed in more detail in the resources below.

How Stuff Works: How a Car Engine Works
The How Stuff Works page offers a complete guide to how gasoline engines work.
http://www.howstuffworks.com/engine.htm

Graedel, Thomas, and Braden Allenby. *Industrial Ecology and the Automobile*. Upper Saddle River, NJ: Prentice Hall, 1998.
The authors use the automobile for an extended case study of how a technological system affects society and the environment. They analyze the flow of energy and materials involved in automobile design, manufacture, use, and infrastructure development.

Cioffi, D. "Physicists Find Challenges in Building Better Cars." *Physics Today*, April 1995, pp. 48, 73-75.
The author explains the distribution of energy input in a typical passenger vehicle and the extent to which energy is dissipated in aerodynamic friction and braking. Engine losses alone account for more than 60 percent of energy used in urban driving.

DeCicco, J., and M. Ross. "Improving Automotive Efficiency." *Scientific American*, December 1994, pp. 52-57.
This article is a more accessible version of the *Physics Today* article above in which the authors explain why only about 12 percent of a car engine's energy output is converted into usable kinetic energy.

Oldham, Scott. "A Century of Cars: A Look Back at the Defining Device of the 20th Century." *Popular Mechanics*.
The author provides a brief history of cars and car culture, including the fact that there have been more popular songs about cars, or songs that refer to cars, than any other topic except love. There's even a song about an engine, the "409." Available at
http://popularmechanics.com/popmech/auto3/0001AUTKP.html

Bleviss, D., and P. Walzer. "Energy for Motor Vehicles." *Scientific American*, September 1990, pp. 103-109.
The authors review development of efficient combustion designs, alternative fuels, and transportation technologies.

Department of Transportation. *Transportation Statistics Annual Report*, 1999.
This comprehensive government report includes statistics on most aspects of transportation including worldwide motor vehicle registration trends, transportation safety, and passenger miles traveled per day. The report notes the dramatic increase in sports utility vehicles and other light trucks, which now account for one-third of the nation's vehicle fleet. Available at *http://www.bts.gov/programs/transtu/tsar/tsar99/tsar99pt.html*

Goddard, Stephen. *Getting There: The Epic Struggle Between Road and Rail in the American Century*. Chicago: University of Chicago Press, 1994.
In this interesting historical account, Goddard chronicles the rise and fall of the railroad. The rapid spread of railroads across the U.S. greatly facilitated nationwide mobility of

goods, but development of a national infrastructure for the automobile made it possible for motor vehicles to supplant trains as the main means of transport.

McPhee, John. "Duty of Care." ***The New Yorker,*** **June 28, 1993, pp. 72-80.**
In this lengthy essay, writer McPhee explores the world's largest pile of scrap tires and reviews the challenge of disposing of millions of automobile tires each year. Dumps used exclusively for tire disposal have been known to catch fire and burn for months. McPhee discusses initiatives to reuse old tires, including waste-to-energy combustion for generating electricity.

Alternative Carbon-Based Fuels

To reduce reliance on petroleum, other carbon-based fuels have been suggested as options, including methanol, ethanol, and compressed natural gas. Any flammable gas, or any solid or liquid that can be converted into a flammable gas, can be used in an internal combustion engine. Ethanol and other fuels were used in early vehicles, particularly during World War II when petroleum supplies were scarce. Responding to concerns about security of oil supplies after the 1990-1991 Persian Gulf War, Congress passed the U.S. Energy Policy Act of 1992, setting a national goal of replacing 30 percent of petroleum in the federal and state fleets of light trucks by the year 2010.

Although fuels such as methanol, ethanol, and natural gas would relieve reliance on petroleum, they are carbon-based fuels, and produce carbon dioxide emissions. Methanol use in automobile engines would slightly increase the incidence and severity of photochemical smog, and methane is also a potent greenhouse gas. Increased use of ethanol would require conversion of additional land to grow the corn to be converted into fuel, and the process of conversion is itself energy intensive. The major impediment to further use of alternative carbon-based fuels, however, is the lack of an infrastructure for refueling. There are also other disadvantages for the consumer, including shorter driving ranges and reduction of space in the vehicle because bulky storage tanks are required.

Fouda, Safaa A. "Liquid Fuels from Natural Gas." ***Scientific American,*** **March 1998, pp. 92-95.**
The author outlines the traditional means of converting natural gas into a liquid and discusses advanced technologies being developed.

Emsley, John. "Energy and Fuels." ***New Scientist*** **141 (January 15, 1994): 1-4.**
This article provides a good overview of the current status of the major non-petroleum-based fuels.

Energy Information Agency. ***Describing Current and Potential Markets for Alternative-Fuel Vehicles.*** **Department of Energy Report DOE/EIA-0604, March 1996.**
In 1992 Congress passed the Energy Policy Act which established goals for increasing the use of alternative fuel vehicles by the federal and state governments. This report from the Department of Energy's Energy Information Agency assesses progress made. Available at *ftp://ftp.eia.doe.gov/pub/pdf/consumption/0604.pdf*

Natural Gas Vehicles 2000
This site is a clearinghouse for information about natural gas vehicles. It includes a basic overview of natural gas vehicle technologies, links to publications, journal articles about natural gas vehicles, a list of refueling stations, and other related information about natural gas. *http://www.ngv.org/*

U.S. Department of Energy, National Biofuels Program: Biofuels for Sustainable Transportation
This government lab maintains a web site containing research updates and descriptions of different types of biofuels. *http://www.biofuels.nrel.gov*

Electric Battery Vehicles

In electric vehicles the power is supplied by a battery rather than by combustion of fuel. The environmental benefit of these vehicles is subject to some dispute, and they have been characterized as "emissions moving" rather than "zero emissions" vehicles. The batteries must be recharged by electricity, and generating the electricity may contribute significant emissions to the atmosphere. Production and disposal of the lead-acid batteries for these vehicles can pose an environmental risk. The most serious impediments to consumer acceptability have been the limited range and performance of these vehicles, together with the cost. Some of the drawbacks of the electric-powered vehicle may be overcome by hybrid vehicles, which use batteries to supply power for cruising but use a gasoline or another type of engine for acceleration. In addition, hybrid vehicles may improve the consumer acceptability of electric vehicles by increasing the distance they may be driven before the batteries must be recharged.

Sperling, Daniel. "The Case for Electric Vehicles." *Scientific American*, November 1996.
This article describes how electric vehicles work and reviews future prospects for this technology. Available at *http://sciam.com/1196issue/1196sperling.html*

Wouk, Victor. "Hybrid Electric Vehicles." *Scientific American*, October 1997.
This author describes the hybrid electric vehicle that uses a small internal combustion engine and an electric generator on board to charge the batteries. He discusses the pros and cons of different electric vehicle configurations and suggests future development plans. Available at *http://www.sciam.com/1097issue/1097wouk.html*

Flavin, Christopher, and Nicholas Lensser. *Power Surge: Guide to the Coming Energy Revolution*. New York: W.W. Norton & Co., 1994, Ch. 10, pp. 193-220.
Flavin and Lensser describe various electric and electric-gas hybrid vehicles that may be potential models for future automobile designs.

Howard, G. "Flat Out for the Car of the Future." *New Scientist* 136 (November 7, 1992): 21-22.
An overview of research related to electric vehicle and battery design.

Fuel Cells

Hydrogen has "long been considered to be the exemplary *energy carrier*, especially for transportation applications,"[12] although it is not a new technology. The hydrogen fuel cell was invented in 1839; and in 1874, a character in the Jules Verne novel, *The Mysterious Island*, suggested that hydrogen would be used as the energy of the future.

The technologies for manufacturing hydrogen are well known. Hydrogen can be produced from water, but the process is expensive. Most designs call for producing hydrogen from carbon-based fuels, such as methane or propane. In a fuel cell, when hydrogen gas meets a catalyst (usually platinum), the catalyst strips electrons from the hydrogen, leaving only hydrogen ions (protons) behind. The protons diffuse through a barrier, a membrane porous enough for the protons but not porous enough to permit the hydrogen molecules to pass through. The electrons flow out of the fuel cell, through an electrode, to an external circuit. After protons pass through the membrane, in the presence of oxygen, they draw the electrons through the circuit, creating a flow of electricity.

The power emitted by a fuel cell, and the reduction of its current high cost, depend on technological advances in the fuel cell membrane, which has to be impermeable to hydrogen and oxygen, yet allow hydrogen ions to pass through. The technology remains enormously expensive. The other technological hurdle that has to be crossed before fuel-cell powered automobiles enter the mass market, is to provide a source of hydrogen. Hydrogen gas is highly flammable (remember the Hindenberg) and huge storage tanks would be required in cars. Gasoline and methanol can be used as sources of hydrogen, which means that traditional tanks and infrastructure (gas stations) could be retained, but these are mixes of hydrocarbons and other impurities, so they would have to be converted or "reformed" before they can be used. Each car would have to contain a "miniature chemical factory" to convert substances into hydrogen. For more on the future of fuel cells, the following resources may be useful:

Kartha, Sivan, and Patrick Grimes. "Fuel Cells: Energy Conversion for the Next Century." ***Physics Today*****, November 1994, pp. 54-61.**
Outlining potential applications as well as the theory behind fuel cells, the authors suggest that the promise of fuel cells remains great because the most important scientific problems for development of fuel cells are in the process of being solved.

"Special Report: Fuel Cells." ***Scientific American*****, July 1999, pp. 72-93.**
This special section includes three separate articles discussing the current status of fuel cell development and its prospects for future use in everything from homes to cars and cell phones.

Beardsley, Tim. "Beyond Batteries." ***Scientific American*****, December 23, 1996.**
The author observes that, even in a world with cheap oil, manufacturers are refining a number of different fuel cell designs for automobiles. Available at
http://www.sciam.com/explorations/122396explorations.html

Service, Robert F. "Bringing Fuel Cells Down to Earth." ***Science*** **285 (July 30, 1999): 682-685.**
Virtually every automobile manufacturer is working on fuel cell technologies but technological hurdles remain. Fuel-cell powered vehicles, for example, would require "miniature chemical factories" on board to produce hydrogen for the fuel cell. This article discusses the design challenges that manufacturers are trying to address.

Electro-chem-technic
This fuel cell manufacturing company's page has educational materials about different types of fuel cells and links to other fuel cell-related web sites.
http://www.i-way.co.uk/~ectechnic/HOME.HTML

National Aeronautics and Space Administration, Kennedy Space Center: Electrical Power System
Electricity on the space shuttle is generated by fuel cells. This page has a detailed description of how the fuel cells generate electricity and how they function within the space shuttle electrical power system.
http://www.ksc.nasa.gov/shuttle/technology/sts-newsref/sts-eps.html

McCosh, Dan. "The Long and Winding Road to Fuel Cells." ***Popular Science*****, March 1997, p. 27.**
Chrysler has developed a fuel cell system that derives hydrogen from gasoline, providing a 30 percent improvement in fuel efficiency over traditional gasoline engines and produces fewer emissions.

U.S. Department of Energy: Hybrid Electric Vehicle Program
The government program's site contains information about fuel cells and links to companies that manufacture fuel cells and fuel cell vehicles. *http://www.hev.doe.gov/*

Simanaitis, Daniel, "Is There an H_2O APU in Your Future?" ***Road and Track*****, November 1999, pp. 114-117.**
The opinion of auto enthusiasts will influence the marketability of future car designs. This *Road and Track* article reviews BMW's new design which uses a fuel cell not to provide "tire-turning torque" (too inefficient), but to supply the energy requirements of the automobile's electrical systems, freeing the engine from these tasks. *Road and Track*'s December 1994 issue also contains a detailed discussion of how automobile manufacturers plan to turn the fuel cell, "an elegant concept," into a practical producer of engine power.

Nuclear Energy

At this point we return to a discussion on energy sources. The remainder of this chapter explores sources used mainly for the purpose of generating electricity, although solar and biomass are also used for domestic heating.

Nuclear power provided approximately 17 percent of the electricity generated worldwide in 1996, and accounted for approximately 6 percent of total worldwide energy consumption.[13] Although nuclear power plants produce no carbon dioxide or other air pollutants, resistance to this energy source has increased because of concerns about its safety and disposal of radioactive waste. Several countries, however, rely on nuclear power extensively to generate electricity. France, for example, generates almost 80 percent of its electricity from nuclear power.[14]

An alternative to nuclear fission is nuclear fusion, which produces far less radioactive waste and uses cheap and plentiful fuel: various forms of hydrogen. With current technology, however, the energy needed to create the extreme conditions in which a fusion reaction can take place exceeds the amount generated in the fusion reaction.

A more promising option is the fast breeder reactor, which converts unfissionable Uranium-239 into usable Plutonium-239, producing a supply of fuel and producing less radioactive nuclear waste. In addition, new reactor designs utilize passive safety systems that depend on natural phenomena and the laws of physics rather than computer or human response to shut down the nuclear reaction if problems occur.

Lawrence Berkeley National Laboratory: ABCs of Nuclear Science
This federal research site provides a basic introduction into the science behind nuclear power, with diagrams and brief explanations of key terms associated with nuclear power and radioactivity. *http://user88.lbl.gov/NSD_docs/abc/home.html*

Wolfson, Richard. ***Nuclear Choices: A Citizen's Guide to Nuclear Technology.*** **Cambridge, MA: MIT Press, 1991, Ch. 1, pp. 1-14; Ch. 2, pp. 15-38.**
Designed for the non-specialist, this book covers the major aspects of nuclear energy: atomic structure, nuclear radiation, electricity from nuclear power, nuclear fusion and nuclear waste, as well as nuclear weapons.

Deutsch, Robert. ***Nuclear Power: A Rational Approach.*** **Columbia, MD: GP Courseware, 1986.**
Deutsch provides a concise guide to nuclear power with brief chapters on the most important areas of concern and instructive diagrams on nuclear power plants and the nuclear fuel cycle, and the licensing process for nuclear plants.

International Atomic Energy Agency
The site of this United Nations intergovernmental forum has data on international nuclear energy use and production. The site also contains articles discussing nuclear power in the context of environmental and development concerns.
http://www.iaea.org/worldatom/

Kabanov, Leonid. "Future Nuclear Power Plants: Harmonizing Safety Objectives." ***IAEA Bulletin*** **37 (1995): 12-15.**
This article is from the quarterly journal of the International Atomic Energy Agency, part of the United Nations system. The author reviews international research efforts to design nuclear reactors that are inherently safe and more resistant to severe accidents. Passive safety features, relying on natural forces such as convection or gravity, reduce the need for human or computer intervention when problems occur.
http://www.iaea.or.at/worldatom/inforesource/bulletin/bull374/kabanov.html

Public Broadcasting Service: Nuclear Reaction—Why Do Americans Fear Nuclear Power?
This public television site contains articles and other documents with scientific and political background information on nuclear power, as well as on the risks of nuclear accidents. *http://www.pbs.org/wgbh/pages/frontline/shows/reaction*

Argonne National Laboratory: Reactor Homepage
This federal laboratory page explains the design of each reactor the lab built, including the passively safe Integral Fast Reactor and its prototype, the Experimental Breeder Reactor II. *http://www.anlw.anl.gov:80/anlw_history/reactors/main.html*

Particle Adventure
The Particle Data group at Lawrence Berkeley National Laboratory developed this guided tour through both the history of nuclear science and the structure of the atom.
http://ParticleAdventure.org/

U. S. Department of Energy, Office of Civilian Radioactive Waste Management. ***Science, Society, and America's Nuclear Waste.*** **Washington, DC: U.S. Department of Energy, 1995.**
This is a four-unit curriculum for secondary-level students covering nuclear waste, ionizing radiation, the Nuclear Waste Policy Act, and the Waste Management System. Each unit contains background information, lesson plans, lab activities, tests, and glossaries. This can be ordered by calling 1-800-225-6972.

Pool, Robert. *Beyond Engineering: How Society Shapes Technology.* (Sloan Technology Series). Oxford: Oxford University, 1997.
The author states that he originally intended to write a history of the nuclear power industry, but as he studied the field, he found the more interesting subject to be the way human factors shape technology. This is a useful study of the interrelationship between technology and society using nuclear power as a case study.

Jaworowski, Zbigniew. "Radiation Risk and Ethics." *Physics Today*, September 1999, pp. 24-29.
In this provocative article, the author argues that the current assumption that radiation at any level is dangerous has led to irrational fear and has imposed enormous costs, reducing the amount available for other life-saving measures. Available at *http://www.riskworld.com/Nreports/1999/jaworowski/NR99aa01.htm*

Nuclear Waste and Security

Holt, Mark. "Civilian Nuclear Waste Disposal." Congressional Research Service Issue Brief for Congress, No. 92059, November 8, 1999.
This report gives a general overview of the issues involved in nuclear waste disposal. *http://www.cnie.org/nle/waste-2.html*

Special Issue: "Radioactive Waste." *Physics Today*, June 1997.
This special issue includes articles that assess the magnitude of the problem of nuclear waste, the technical challenges, and the potential of future strategies for managing and disposing of waste.

Stover, Dawn. "The Nuclear Legacy." *Popular Science*, August 1995, pp. 52-58, 81, 83.
This article points out that because the United States has not resolved conflicts surrounding the storage of spent nuclear fuel rods, the rods are usually stored in cooling ponds located at the power plants themselves.

Williams, Phil, and Paul Woessner. "The Real Threat of Nuclear Smuggling." *Scientific American*, January 1996, pp. 40-44.
According to this article, nuclear materials stored at military facilities in the former Soviet Union are poorly supervised and are a tempting target for smugglers, including rogue nations, terrorists, and crime groups.

Yucca Mountain Project
The U.S. Department of Energy hopes to place a nuclear waste disposal facility at Yucca Mountain, Nevada. This project site contains fact sheets and historical information about nuclear waste disposal in general and the Yucca Mountain facility in particular. *http://www.ymp.gov/about/index.htm*

Radwaste.Org
This page links to sites relating to radioactive waste, including nuclear and radiation-related companies, research centers, regulatory agencies, and both government and non-governmental organizations. *http://www.radwaste.org*

Alternative Energy Technologies

Many alternative energy sources, such as wind and biomass, have been used for thousands of years. The term "renewable energy" is generally used to describe energy derived from the Sun, wind, and water, and sources that are available continuously over shorter time frames than are supplies of petroleum and coal resources, which are replenished only over many millions of years. A crucial problem for renewable energy is whether intermittent energy sources such as solar and wind can contribute reliably to the base load of electricity demand that electrical utilities must provide to their consumers. With traditional sources such as coal-powered, nuclear, or hydroelectric power plants, utilities can generate electricity when it is needed by simply turning a switch to start the turbines. Renewable technologies such as photovoltaic cells and wind turbines can provide electricity only when the Sun shines or the wind blows, not necessarily when consumers demand it. There is currently no practical technology for large-scale storage of electricity generated by intermittent sources such as wind and solar.

National Renewable Energy Laboratory
The site of this government research lab links to pages offering statistics, fact sheets, and legislative documents pertaining to renewable energy, as well as a detailed page entitled "Clean Energy 101." *http://www.nrel.gov*

U.S. Department of Energy: Energy Efficiency and Renewable Energy Network
This site links to other departments, offices, and laboratories within the United States Government that are devoted to research and development of renewable and alternative energy sources. *http://www.eren.doe.gov*

Solstice
Solstice is an online clearinghouse that provides links to sites with information on renewable energy resources and technology. *http://solstice.crest.org/renewables/index.shtml*

U.S. Energy Information Agency: Hydroelectricity and Other Renewable Resources
This federal government site includes a statistical review of electricity production from renewable sources. *http://www.eia.doe.gov/oiaf/ieo97/hydro.html*

Solar Energy

We often speak of solar energy as distinct from the other forms of energy, but most of the energy we use originated as solar energy. Fossil fuels contain the energy that plants absorbed from the Sun eons ago, and the biomass burned in many places contains solar energy stored more recently. Windmills are driven by air masses pushed around the troposphere by uneven solar heating. Even if they are not equipped with solar panels, our homes absorb enough solar energy—directly or indirectly—to keep our furnaces off for long stretches of time. But there are vast differences between passively absorbing thermal energy from the Sun to heat our homes and converting that energy into more versatile forms such as electricity. These are differences of technology and cost.

The Sun's energy can be exploited in a number of ways. Passive solar heating is the simplest method of harnessing solar energy. Passive solar designs reduce the heating and cooling requirements of a building through siting as well as through using windows and other materials to collect and hold heat. Active solar systems, on the other hand, usually collect the Sun's energy with large panels though which air or liquid is pumped into a storage area and then recirculated to the panels. The thermal heat of the Sun can also be converted into electricity through a photovoltaic cell. Photovoltaic cell technologies remain costly, which has limited the competitiveness of this source. In remote areas, however, where electricity is not available from other sources, photovoltaic cells may be the best option.

U.S. Department of Energy: Photovoltaics Program
This federal government site explains how solar cells work, what they are made of, and how they are used. *http://www.eren.doe.gov/pv/*

Hoagland, William. "Solar Energy." *Scientific American*, September 1995, pp. 136-139.
The author provides background statistics on the world's daily solar energy input and discusses the solar thermal and photovoltaic technologies. Wind energy, he reminds us, is also a type of solar energy.

Shah, A. et al. "Photovoltaic Technology: The Case for Thin-Film Solar Cells." *Science* 285 (July 30, 1999): 692-698.
This is a technical, detailed overview of current photovoltaic technologies and the prospect of developing materials that would make the cells less costly than they currently are.

Stone, Jack. "Photovoltaics: Unlimited Energy from the Sun." *Physics Today*, September 1993, pp. 22-29.
The author uses a basic discussion of the physics and, to a lesser extent, the economics of photovoltaics, to argue that it is an increasingly viable energy technology.

Williams, Neville, Ken Jacobson, and Harold Burris. "Sunshine for Light in the Night." *Nature* 362 (April 22, 1993): 691-692.
The authors argue that photovoltaic-solar-power systems are the best option for bringing electricity to isolated rural areas in developing countries. Although rural markets rarely have the economy-of-scale necessary to attract investment in power lines, small private companies or cooperative ventures are using small solar-powered systems in some areas to generate electricity.

Gunther, Judith. "Power Houses." *Popular Science*, May 1996, pp. 77-81.
A number of companies are bringing hopes for cost-effective photovoltaic systems closer to reality using crystalline and thin film cells, and even transparent "nanocrystalline" cells, that can double as windows.

Whitney, Hunter. "When Science Imitates Nature." *Omni*, May 1994, p. 25.
Researchers are working toward a new type of solar energy that harvests the Sun's rays by mimicking the process of photosynthesis.

Spinks, Peter. "Plug into the Sun." *New Scientist* 127 (September 22, 1990): 48-51.
The first photovoltaic cell, which converted sunlight into electricity, was made in 1876, but this technology remains expensive and inefficient. This article suggests that it is more efficient to use solar energy directly to heat buildings than to convert it into electricity which is then used to power heating equipment. Thermal heating accounts for more than 30 percent of global energy use; more reliance on passive solar systems would reduce the use of primary energy used for domestic heating.

Hydroelectric Power

Hydroelectric energy is the kinetic energy of flowing water captured by dams, channeled through turbines, and converted into electrical energy. Hydropower is a relatively clean energy source, producing electricity without the emissions associated with power plants using fossil fuels. Dams, however, alter and sometimes destroy the surrounding ecosystems. Adverse impacts on salmon populations in the West, for example, have led to increased opposition to dams in that region (for more on dams, see Water).

International Council on Local Environmental Initiatives: Energy Fact Sheet
Originally published by the Energy Educators of Ontario, this brief introductory page on hydroelectric power discusses the history, types, and environmental implications of hydropower. *http://www.iclei.org/efacts/hydroele.htm*

National Renewable Energy Laboratory: Hydroelectric Power
This government research site provides an overview of hydroelectric power, including history, types, and environmental impacts.
http://www.nrel.gov/lab/pao/hydroelectric.html

Foundation for Water and Energy Education
This consortium of government agencies, research organizations, and electric utilities seeking to provide information on water and hydropower issues. This site contains some basic lessons on hydropower and links to sites with science and hydropower teaching resources. *http://www.fwee.org*

National Geographic Geoguide: Dams
This educational site has brief discussions of hydroelectricity as well as an online tour of a river before, during, and after construction of a dam.
http://www.nationalgeographic.com/resources/ngo/education/geoguide/dams/splashframe.html

Wind

Wind turbines were first used to generate electrical power in Denmark a century ago, and centuries before that windmills began providing the power for milling grain and watering crops. During the years from 1850 to 1970, over six million windmills and wind turbines were built throughout the Midwestern United States for pumping water on farms and in rural areas (Righter, 1996). In recent years, wind power has been the subject of renewed interest because it is a relatively clean, low-cost source of electrical power for those areas with sufficient wind. The Netherlands, for example, has a program to subsidize the construction of new windmills and hopes to use wind power for a significant portion of its energy needs. California is expanding its capacity to generate electricity from this source. The large wind farms that are required to generate electricity, however, are noisy and pose significant threats to birds. The development of reliable equipment is also a challenge; blade failure is a common problem.

Look there, my friend Sancho Panza, where thirty or more monstrous giants present themselves, all of whom I mean to engage in battle and slay! [Don Quixote said, pointing to the windmills dotting the Spanish landscape.]

★Miguel De Cervantes, **Don Quixote**★

U.S. Department of Energy: Wind Energy Program
This federal government site includes papers, photos, data, and brief fact sheets on wind power production, research, and resources in the United States.
http://www.eren.doe.gov/wind

Renewable Resource Data Center: Wind Energy Resource Atlas of the United States
This federal government site offers wind resource maps and describes potential wind energy production for the regions of the United States.
http://rredc.nrel.gov/wind/pubs/atlas/

Righter, Robert. *Wind Energy in America*. Norman, OK: University of Oklahoma Press, 1996.
Righter provides a history of wind power and wind technology in the United States from colonial times to the present.

American Wind Energy Association
This organization is devoted to expanding wind energy use in America. The site includes FAQ pages on wind energy technology and links to order educational resources.
http://www.awea.org/

California Energy Commission: Wind Energy in California
This state site includes maps and fact sheets about the renewable energy possibilities in California, which produces more electricity from wind turbines than any other state.
http://www.energy.ca.gov/wind

Gantenbein, Douglas. "Something in the Wind." *The Atlantic Monthly,* October 1993, pp. 36-42.
The author notes that, with its low population density and high average winds, and the development of new wind turbines, South Dakota could become "the Saudi Arabia of airflow."

Wind Service Holland
This organization's site contains a number of pages on use of wind energy in the Netherlands, a pioneer in the use of windmills, and includes a page of statistics on the country's installed capacity for wind energy. *http://home.wxs.nl/~windsh/english.html*

National Renewable Energy Laboratory: National Wind Technology Center
This government research site describes the workings of different wind turbine designs and discusses environmental issues related to the use of wind energy, such as "avian issues" (i.e. the safety of birds flying near wind turbine blades). *http://www.nrel.gov/wind*

Stover, Dawn. "The Forecast for Wind Power." *Popular Science*, July 1995, pp. 66-72.
As interest in wind power increases, a number of wind turbine designs are competing for attention and markets. The economics of energy, however, make near-term prospects for wind energy uncertain.

Clarke, Alexi. "How Green is the Wind." *New Scientist* 122 (May 27, 1989): 62-65.
Britain is planning a significant investment in wind energy farms, but as this article reports, there are environmental impacts and engineering considerations that must be addressed. Among other problems, planners in Britain are having difficulty finding suitable land for the windmill parks because many consider them unsightly, noisy, and unsafe.

Biomass

Until the Industrial Revolution, biomass was the main source of energy for human beings and for many parts of the world, it remains the primary source for much of the population. Firewood, the remains of crops, dung, and other matter provide heat for cooking, warmth, and for small-scale manufacturing. Biomass is considered a renewable resource because crops, trees, and other sources can be replanted, but management is required to ensure continued renewability – in a number of countries where biomass provides the only source of fuel, serious deforestation has occurred. Interest in biomass as an alternative to fossil fuels began several decades ago, and technologies for converting biomass into gaseous and liquid fuels are still being developed. The environmental impacts of this source, though, would be substantial, with wider use. Large areas of agricultural land would have to be converted to growing biomass materials, placing further stress on ecosystems. Biomass combustion is a source of hydrocarbons and air pollutants, and the offensive odors emitted from biomass conversion plants would make placing these plants difficult.

National Renewable Energy Laboratory: Biomass Resource Information Clearinghouse
This government research page links to basic fact sheets, data, and bibliographies on biomass and bio-energy. *http://rredc.nrel.gov/biomass/*

Energy Efficiency and Renewable Energy Network: Biopower Program
This federal government page discusses a number of technologies for extracting the energy stored inside plant biomass, such as combustion, co-firing, and gasification.
http://www.eren.doe.gov/biopower/

Geothermal

The heat produced by the decay of radioactive substances in the Earth's core is known as geothermal energy. The heat spreads through the interior of the Earth, heating the fluids underneath the surface. These fluids escape to the surface in areas where there are fissures and cracks in the crust, sometimes in the form of geysers or volcanoes. Old Faithful in Yellowstone National Park is a well-known example of geothermal energy. Heat from geothermal vents is exploited in several ways. One method used to generate electricity is a steam-turbine plant. Water or other medium is pumped into an injection well (which may need to be as deep as 10,000 feet), which brings the hot water to the surface. These fluids are under high pressure, and when the pressure is released at the surface (in vessels known as flash tanks), and a portion of the water (about 35 percent depending upon its temperature) flashes (boils) to steam. The steam is separated from the remaining water, and the steam is piped to a turbine engine, which turns a generator.

Because of the high temperature and corrosive nature of geothermal fluids, and the hard rocks found in geothermal environments, geothermal drilling is difficult and expensive. Geothermal plants also produce significant emissions of carbon dioxide, in addition to hydrogen sulfide, mercury, boron, arsenic, and radon. Geothermal vents are also often located in pristine areas (such as Yellowstone), so construction of plants in these areas would pose additional environmental costs. Moreover, geothermal energy is not widely distributed. The Geothermal Power Association estimates that at best, geothermal energy could supply about 5 percent of U.S. energy needs.

National Renewable Energy Laboratory: Geothermal Energy
This government research page provides basic information on geothermal energy.
http://www.nrel.gov/lab/pao/geothermal_energy.html

Energy and Geosciences Institute, University of Utah: Geothermal Energy
This university research institute site provides a basic introduction to geothermal energy with diagrams, a video, and explanations of how geothermal sources are tapped and converted into usable energy. *http://www.egi.utah.edu/geothermal/geothermal.htm*

U.S. Department of Energy, Renewable Energy: Geothermal Energy
This federal government site offers information primarily on U.S. use of geothermal energy. *http://www.eren.doe.gov/consumerinfo/refbriefs/tpgeo.html*

U.S. Geological Survey. *Tapping the Earth's Natural Heat.* U.S. Geological Survey Circular 1125. Washington, DC: U.S. Geological Survey, 1997.
This booklet provides an overview of geothermal energy, including what is known about geothermal use in ancient times, how it is used today, and how it will likely be used. This report can be ordered from 1-888-ASK-USGS.

CHAPTER 7 Forests

Background

Forested lands are areas of great natural beauty and biological diversity. They also provide essential natural resources. Humans have always relied on forests for shade, shelter, and warmth (firewood), and building materials. Forests are a renewable resource, if properly managed. However, throughout history, humans have changed forests through fire and by clearing land for agriculture.

> First, what is forestry? Forestry is the knowledge of the forest. In particular, it is the art of handling the forest so that it will render whatever service is required of it without being impoverished or destroyed...A forest is not a mere collection of individual trees, just as a city is not a mere collection of unrelated men and women...A forest, like a city is a complex community with a life of its own. It has a soil and an atmosphere of its own, chemically and physically different from any other, with plants and shrubs as well as trees which are peculiar to it. It has a resident population of insects and higher animals entirely distinct from that outside.
>
> ★Gifford Pinchot, **The Training of a Forester★**

What are the issues?

Significant clearing of temperate forests occurred in Europe and North America during the 17th and 18th centuries. England's forests were devastated by its need for timber for shipbuilding, fuel, and other uses. One man-of-war, for example, required timber from many thousands of trees.[15] Development of alternate energy sources, primarily coal and oil, and of building materials such as iron, relieved the pressure on European and American forests. High yield farming methods developed in the latter half of the twentieth century have permitted the regeneration of forests on many acres of land previously used for agricultural purposes. Most commercial logging in the U.S. now occurs on tree plantations grown for that purpose and much of the remaining old growth forest in the United States is in protected areas, although debates continue about access to timber on federal lands.

Concerns about deforestation are most serious in developing countries. Wood cutting in sub-Saharan Africa has contributed to the spread of desertification, and some developing countries are still clearing tropical forests for agriculture and other uses at a rapid rate. For much of the population in undeveloped countries, especially Africa, firewood remains the major source of energy.

What are the uncertainties?

Estimates of global rates of deforestation vary widely. It is thought that the most rapid tropical forest loss has occurred in Brazil, Mexico, Venezuela, Indonesia, Zaire, Bolivia, Thailand, Philippines, El Salvador, Honduras, and Paraguay, but accurate data are difficult to obtain. Satellite observations are now being used to develop more accurate data about forest cover. There is also some disagreement concerning the causes of desertification (the spread of arid, desert-like conditions) and its extent. Overgrazing of land, wood collecting, and other human activities are thought to be important causes, but the role of climate changes is not known with certainty.

> The tendency nowadays to wander in wilderness is delightful to see. Thousands of tired, nerve-shaken, over-civilized people are beginning to find out that going to the mountains is going home; that is a necessity; and that mountain parks and reservations are useful not only as foundations of timber and irrigating rivers, but as fountains of life.
>
> ★John Muir, **Our National Parks**★

Forest re-growth in the United States and in Europe has been significant. There have been some areas of forest decline, and it was thought that damage was due to acidification of soils from acidic precipitation. It is now theorized that "die back" conditions may be largely due to climate factors and forest health has improved in some areas thought to have been adversely affected. The forest resources of Europe experienced their most rapid recorded regrowth during the decade of the 1980s, growing at a rate about thirty-five percent higher than in earlier decades.[16]

What are the Risks and Tradeoffs?

Forests provide many services, including serving as major stores of carbon. Deforestation, especially clearing land by burning, releases large amounts of carbon dioxide, a greenhouse gas, into the atmosphere. Tropical forests are the most biologically diverse areas on the planet, so clearing forests in these areas puts more species at risk than in temperate areas. However, in undeveloped countries forests are cleared to meet essential human needs. International concerns about forest loss have little effect when populations in these countries have few options for survival. Efforts to find alternative sources of income from forest lands, such as pharmaceuticals or other marketable products, have not yet provided sufficient incentive to prevent deforestation in many areas.

Basic Concepts

The science of forestry, called silviculture (from the Latin word, *silva*, meaning forest), has a long history, dating to ancient Roman times (Chase, 25). The modern discipline began in Europe in the mid-1600s. Charles II of England, fearing England would run out of trees to build war ships, encouraged the Royal Society to study the management of the country's timber reserves. Gifford Pinchot, the first director of the Forest Service in the United States, went to Europe to study the scientific management of forests.

There are college-level forestry science textbooks for detailed study; many of the basic concepts related to the dynamics of a forest ecosystem are covered in upper level biology or ecology textbooks, including the differences between various types of forest, the flow of nutrients through the forest ecosystem, and how forests react to change and disturbances. The following resources may be helpful.

Smith, Thomas, and Robert Smith. *Elements of Ecology*. Menlo Park, CA: Addison Wesley Longman, Inc., 1998.
This textbook provides a basic introduction to forest ecosystems, explaining the physical, geographical, climatic, and other biological factors that differentiate the various types of temperate and tropical forests.

Botkin, Daniel B. *Dischordant Harmonies*. Oxford: Oxford University Press, 1999.
Botkin illustrates the dynamic equilibrium of natural ecosystems, including forests, and provides a clear explanation of forest succession, uses of clearcutting, and ecosystem dynamics.

deBlij, H., and P. Muller. *Physical Geography of the Global Environment*. New York: John Wiley & Sons, Inc., 1998.
The textbook provides a basic introduction (with extensive maps) of the physical and geographic factors that delineate biomes and different types of forest land.

Sierra Nevada Ecosystem Study
This research project site posts the current findings of a three-year study of the ecosystems of the California Sierra Nevada Range.
http://ceres.ca.gov/snep/pubs/web/v1/v1_default.html

Luoma, Jon. *The Hidden Forest: A Biography of an Ecosystem*. New York: Henry Holt and Co., 1999.
Luoma describes the research of scientists in an old-growth forest ecosystem, Andrews Experimental Forest in Oregon. This is a long-term, interdisciplinary study in which botanists, ecologists, soil biologists, and entomologists are working together to study the roles of every element in a forest ecosystem.

Heinrich, Bernd. *The Trees in My Forest*. New York: Cliff Street Books, 1997.
The author, a biologist, combines personal memories of the forest, drawings, and rich descriptions of the flora and fauna, with a detailed introduction to forest ecology. He illustrates, for example, the factors that determine a tree's shape, how trees are distributed, and the symbiotic relationship between ants and trees.

Williams, Michael. *Americans and Their Forests: A Historical Geography*. Oxford: Cambridge University Press, 1989.
This is an extensive study of American forests from pre-European times to contemporary times. Williams traces the clearing of forests that took place in the 17th and 18th centuries and the subsequent regeneration in decades since World War II.

Wessels, Tom. *Reading the Forested Landscape: A Natural History of New England*. Woodstock, VT: The Countryman Press, 1997.
The author uses the forests in his native New England to demonstrate how to "read the landscape." He shows how to find evidence of human activities and disturbances such as fire, logging, pastures, and blight, to trace the history of the land and the forests.

Muir, John. *The Mountains of California*. New York: Penguin Books, 1985.
This reprint of a book by one of the founders of the American conservation movement, includes detailed descriptions of the wilderness of California's Sierra range. In Chapter VIII, "The Forests," Muir discusses individual tree species and the forest environment from both scientific and aesthetic perspectives.

National Geographic Habitats: The Growth of a Forest
This site depicts a West Virginia forest at different stages of growth: 2 years, 12 years, 41 years, and 86 years. *http://www.nationalgeographic.com/habitats/index2.html*

MacCleery, Douglas. *American Forests: A History of Resiliency and Recovery.* Durham, NC: Forest History Society, 1994.
The author documents the changes in American forests from colonial times to the present, and includes data about the extent of U.S. forests prior to European settlement, clearing of forests by Native Americans and early settlers, and regeneration that has occurred over the last few decades. He reports that today there is more forestland in the U.S. than there was 75 years ago.

Data

U.S. Forest Service
This federal agency provides information on forest health in the United States, maps of U.S. forest growth, and information about current forest policies. *http://www.fs.fed.us*

United Nations Food and Agriculture Organization: Forestry
This international agency is considered to be one of the most authoritative source of data on the world's forests. Their publication entitled "The State of the World's Forests" is available for download in *.pdf* format.
http://www.fao.org/WAICENT/FAOINFO/FORESTRY/forestry.htm

ForestWorld
This industry site provides basic information covering the extent of forested land, and a brief background of the geography, language, and climate of countries around the world.
http://www.forestworld.com/

World Resources Institute: Frontier Forests Initiative
Russia contains the Earth's largest expanse of frontier forests. This project site tracks the world's remaining frontier forests, and includes extensive maps of current and historical forest cover. *http://www.wri.org/ffi/maps/*

Woods Hole Research Center: World Forest Resources
This research center has constructed maps from data acquired from satellites of forested land in the tropics of South America and in the former Soviet Union.
http://www.whrc.org/science/Globfor/globfor.htm

Forest Management: United States

In the 19th century recognition grew that the nation's forest reserves were being depleted and that conservation measures were necessary. National forest reserves were created in the United States in 1891 and the U.S. Forest Service was created in 1905. The mission of the Forest Service was to incorporate methods of scientific forestry, or silviculture, to managing U.S. forest reserves (see A Historical Perspective). Current forestry policy in the United States is divided between the historical approach focused on achieving a steady timber yield ("sustained yield") and a new strategy of ecosystem management.

Alverson, William, Walter Kuhlmann, and Donald Waller. *Wild Forests: Conservation Biology and Public Policy.* Washington, DC: Island Press, 1994.
The authors provide a brief history of forest resource use and forest policy in the United States. They discuss current issues in the ecology of forest management and evaluate current forestry practices, emphasizing the importance of biodiversity.

Chase, Alston. *In a Dark Wood: The Fight over Forests and the Rising Tyranny of Ecology.* Boston: Houghton Mifflin, 1995.
Chase chronicles the political, scientific, and philosophical controversies that have animated disputes over the best way to manage forested lands.

Noble, Ian, and Rodolfo Dirzo. "Forests as Human-Dominated Ecosystems." *Science* 277 (July 25, 1997): 522-525.
Noble and Dirzo survey forest management techniques from "slash-and-burn" forest clearing to intensive logging. They also examine how commercial forestry has broadened its "sustainable yield" approach to take into account biodiversity and ecosystem maintenance.

Report of the Forest Ecosystem Management Assessment Team. *Forest Ecosystem Management: An Ecological, Economic, and Social Assessment.* Washington, DC: Government Printing Office, 1993.
This report, referred to as the FEMAT report, established the policy of ecosystem management for the national forest reserves.

Resources for the Future: Forestry
This research organization provides online access to a number of articles on forest management, forestry economics, and U.S. government forestry policy.
http://www.rff.org/nat_resources/forests.htm

National Council for Science and the Environment: Forestry Issues
These Congressional Research Service briefs address political issues related to U.S. forest lands, but also provide basic overviews of forest status and health.
http://www.cnie.org/nle/crsfor.html

Backiel, Adela, and Ross W. Gorte, "Clearcutting in the National Forests." Congressional Research Service Report for Congress, No. 92-601 ENR, July 29, 1992.
This report for Congress and its staff reviews the controversy over clearcutting on public lands, which is regarded by many as harmful, although foresters argue that, properly managed, it is the appropriate method for harvesting and regenerating certain commercial tree species. This report reviews the advantages and disadvantages of this forest management method. Available at *http://www.cnie.org/nle/for-2.html*

Clark, Earl. "How to Cut Your Forest and Save It Too." *American Forests* 100 (September 1994): 40.
Earl reviews the timber management techniques on Vancouver Island, British Colombia, using selective cutting, which opens up the forest canopy, allowing sunlight to regenerate the forest floor.

Muir, John. ***Our National Parks.*** **Boston: Houghton Mifflin, 1901. Madison, WI: University of Wisconsin Press, 1981 (Reprint).**
This is a collection of ten of Muir's essays written to introduce Americans to the beauty and wonder of their national parks.

Pinchot, Gifford. ***The Fight for Conservation.*** **New York: Doubleday, 1910.**
In this collection of essays, Pinchot explains his belief that natural resources should be scientifically managed to prevent waste and preserve them for the common good.

Sedjo, Roger. "The Global Environmental Effect of Local Logging Cutbacks. ***Resources,*** **Fall 1994.**
Sedjo points out that the price of reducing logging in the Pacific Northwest is that logging occurs elsewhere to meet the need for timber, including in developing countries where forestry may not be as well managed as in the U.S. Ordering information is available at *http://www.rff.org*

Sellars, Richard W. ***Preserving Nature in National Parks.*** **New Haven: Yale University Press, 1997.**
Sellars, a National Park Service historian, provides a detailed account of how the American national parks system was created and how it has evolved. He describes the tensions that have arisen as the Park Service tries to reconcile sometimes-competing management goals including promoting recreation and use of the parks while preserving natural resources.

GLOBAL CONSERVATION STRATEGIES

Deininger, Klaus, and Bart Minten. "Poverty, Policies, and Deforestation: The Case of Mexico." ***Economic Development and Cultural Change*** **47 (January 1999): 313-344.**
This study found that government subsidies and import restrictions distort timber prices and exacerbate forest loss. This article may be found through an electronic periodical database in a public library.

Moffat, Anne. "Resurgent Forests can be Greenhouse Gas Sponges." ***Science*** **277 (July 18, 1997): 315-316.**
This article discusses proposals to plant trees to offset anthropogenic carbon dioxide (CO_2) emissions in light of new research suggesting that the amount of CO_2 absorbed by forests has been underestimated.

Ciais, P. et al. "A Large Northern Hemisphere Terrestrial CO_2 Sink Indicated by the 13C/12C Ratio of Atmospheric CO_2." ***Science*** **269 (August 25, 1995): 1098-1102.**
Approximately half of all carbon dioxide emissions in the Northern Hemisphere do not remain in the atmosphere, but there has not been sufficient data available to assess the amount of this carbon dioxide that is absorbed by the oceans and by the terrestrial biosphere. This article (a challenging one) describes research that suggests a significant amount of carbon dioxide is taken up by land in the Northern Hemisphere, perhaps because of forest regeneration that has occurred in the last few decades.

Raloff, Janet. "Bombs Away!" *Science News*, January 11, 1997, pp. 28-29.
This article describes an effort to release tree seeds from airplanes onto exposed or deforested areas. This approach is intended to increase CO_2 absorption by trees, but the article also discusses the possibility of using this method in broader reforestation programs.

RAINFORESTS

Tropical forest areas cover only a small percentage of the Earth's surface, but they are unusually rich in biological diversity, especially insects and flowering plants (See Biodiversity). These forests lie in countries that are experiencing rapid population growth; and land-intensive farming methods, insecure rights to land, and practices such as burning vegetation to clear land for agriculture have exacerbated forest loss.

Cartoon by Scott Willis of the San Jose Mercury News.

Whitmore, Timothy. *An Introduction to Tropical Rain Forests*. New York: Oxford University Press, 1990.
Whitmore reviews the geologic and ecological history of tropical rain forests, including the distinguishing features of tropical rain forests of Central and South America, Asia, and Africa.

Living Earth: The Rainforest Database
This database includes text and images collected from numerous books and articles. The site gives an overview of rainforest ecology, including the climatic factors that delineate tropical areas, the different types of tropical forests, and an explanation of theories about why the tropics have so much species diversity.
http://www.gn.apc.org/LivingEarth/RainforestDBintro.html

Centre for International Forestry Research (CIFOR)
CIFOR is a partnership of international organizations whose mission is applied research in tropical forestry and the transfer of sustainable forest methods to developing countries. *http://www.cifor.org*

Skole, D. et al. "Physical and Human Dimensions of Deforestation in Amazonia." *BioScience* 44 (May 1994): 314-322.
This article examines the economic and demographic factors related to forest loss in the Amazon basin, noting that agricultural expansion plays the largest part.

Taylor, David. "Saving the Forest for the Trees: Alternative Products from Woodlands." *Environment*, January/February 1997, pp. 6-11, 33-36.
Taylor surveys the production and trade of products from tropical and sub-tropical forests. He argues that measures such as securing property rights and increasing the availability of small-scale credit are needed to minimize harm to forest ecosystems.

Allen, William H. "Traveling across the Treetops." ***BioScience*** **46 (December 1996): 796-799.**
This article describes the Smithsonian Tropical Research Institute's crane, which allows researchers to study the rarely seen ecosystems of tropical forest canopies.

Moffett, Mark. "Climbing an Ecological Frontier: Tree Giants of North America." ***National Geographic*****, January 1997, pp. 46-61.**
Moffett describes research on rain forest canopies, and describes some of the ecological surprises that have been encountered hundreds of feet off the ground.

Shoonmaker, Peter, Bettina von Hagen, and Edward Wolf, eds. ***The Rain Forests of Home: Profile of an American Bioregion*****. Washington, DC: Island Press, 1997.**
This is a collection of studies and essays on the coastal temperate rain forests of Oregon, Washington, British Columbia, and Alaska.

Forest Fires

Fire is the oldest method humans have used to clear land for farming and other uses, and it is still used today in many developing countries. In the past several years, some of these fires have burned out of control, destroying many acres of prime forestland. This is a concern not only because of the threat to biodiversity in those areas, but because deforestation, especially by fire, is an important source of anthropogenic carbon dioxide, contributing around 15 percent of total yearly anthropogenic carbon emissions.

Historically, in U.S. national parks and other areas, when fires from natural or other causes began, efforts were made to control them as quickly as possible. That policy has changed as more has been learned about the role of fire as a natural disturbance in forest ecosystems.

Pyne, Stephen. ***World Fire: The Culture of Fire on Earth*****. New York: Henry Holt and Co., 1995.**
Pyne relates how, at different places and times, humans have used fire to shape their environments and how natural fire has shaped human culture. This is an expensive volume; it may be available at a public or university library.

Romme, William, and Don Despain. "The Yellowstone Fires." ***Scientific American*****, November 1989, pp. 37-46.**
Prior to the 1970s, officials at Yellowstone National Park attempted to extinguish all forest fires. After discovering the benefits of fire on forest ecology, park officials allowed naturally-occurring fires to burn under close surveillance. This article details the various ecological stages that the Yellowstone forest undergoes as a result of fires.

Gorte, Ross. "Forest Fires and Forest Health." Congressional Research Service Report for Congress, No. 95-511 ENR, July 14, 1995.
This report for Congress is a good introductory discussion of both the relationship between forest fires and forest health and the policy options forest officials must consider when fires threaten habitat and timber as well as human settlements.
http://www.cnie.org/nle/for-5.html

Abrams, Marc. "Fire and the Development of Oak Forests." ***BioScience*** **42 (May 1992): 346-353.**
The author traces the development of oak forests in the eastern U.S. over the last 18,000 years, attributing their distribution and variety to the varying influence of fire on the biome.

Discovery Channel Online: Line of Fire—New Respect for an Old Enemy
This television and magazine site includes a number of articles about forest fires, such as "Stressed Out from Being Fire-Free." Smoke jumper stories, current fire maps, and a fire control simulation are also available on the site.
http://www.discovery.com/area/science/wildfires/previous.html

Fire Weather Education
This University of Wisconsin site discusses the relationship between weather and forest fires and includes resources intended to be used in K-12 fire education curricula. The site includes case studies on a handful of real-life wildfire stories, monthly fire weather forecasts, and incidence statistics. *http://earthlab.meteor.wisc.edu/firewx/index.html*

LEGISLATION

National Forest Management Act (16 U.S.C. sec. 1601)
This statute calls for the development of a National Renewable Resource Assessment every ten years to anticipate demand and supply for the country's renewable resources. The Act also requires interdisciplinary, scientific development of land and resource management plans for each National Forest.
http://www4.law.cornell.edu/uscode/unframed/16/ch2.html

CHAPTER 8

Food and Agriculture

Background

Around 10,000 years ago, humans began to domesticate plants and animals. Cultivating crops kept humans together in one place, provided a more secure source of food than hunting and gathering, and enabled some to engage in activities not directly related to subsistence. In short, civilization began—together with the significant human impact on the environment that is its inevitable companion.

The unprecedented growth in global population that occurred in the twentieth century was made possible by remarkable advances in agriculture and technology. The Green Revolution brought high-yield crops and advanced growing techniques to developing countries and improved nutrition and health for most parts of the world. Continuing population growth will require more acres of land to plant and more water for irrigation, increasing pressure on habitats and natural resources.

What are the issues?

In developing countries, where the population growth is most rapid, forests and other previously unmanaged lands are being converted to agricultural use. The negative environmental impact is exacerbated in some areas by poor agricultural practices, including "slash and burn" clearing of land, and depletion of soil minerals from poorly managed farming on marginal land.

Advanced agricultural methods have made it possible to grow more food on fewer acres of land, which permits much land to be returned to forest and other uses. However, there are costs associated with high-yield methods, including their heavy reliance on pesticides and fertilizers. Runoff of fertilizers from agricultural lands has affected the water quality of some lakes, rivers, and bays. Erosion is, and has been since crops were first planted, a serious environmental problem in the absence of preventative measures. Irrigation, required in many parts of the world, can be a significant drain on water supplies in arid areas and, if improperly managed, can lead to buildup of salt deposits, which may severely degrade soil quality.

What are the uncertainties?

One of the most critical challenges of the next few decades is to find ways to increase production of food while minimizing environmental degradation. Considerable research in this and other countries has led to advanced methods to preserve soil and prevent erosion, including conservation tilling and computer-controlled application of pesticides, fertilizers, and water. Other methods are more controversial. To reduce reliance on pesticides, methods such as biocontrol (using other species to control pests) and genetic engineering to create pest-resistant crops are being tested. Concerns about the potential impacts of these methods have been raised.

What are the risks and tradeoffs?

> The purity of foods was as much a concern in the early 20th century as it is now at the beginning of the 21st. At the annual meeting of the National Association of State Dairy and Food Departments in 1903, a report on "The Use of Coloring Matter and Antiseptics in Food Products" was read by James H. Shepard, the State Analyst of South Dakota. This menu was his answer to some food manufacturers who claimed the amount in the portion for a single meal was so small as to be harmless:
>
> BREAKFAST
> Sausage, coal-tar dye and borax
> Bread, alum
> Butter, coal-tar dye
> Canned cherries, coal-tar dye and salicylic acid
> Pancakes, alum
> Syrup, sodium sulphite
> (This gives eight doses of chemicals of dyes for breakfast)
>
> DINNER
> Tomato soup, coal-tar dye and benzoic acid
> Cabbage and corned beef, saltpetre
> Canned scallops, sulphurous acid and formaldehyde
> Canned peas, salicylic acid
> Catsup, coal-tar dye and bezoic acid
> Vinegar, coal-tar dye
> Bread and butter, alum and coal-tar dye
> Mince pie, boracic acid
> Pickles, copperas, sodium sulphite, and salicylic acid
> Lemon ice cream, methyl alcohol
> (This gives sixteen doses for dinner.)
>
> SUPPER
> Bread and butter, alum and coal-tar dye
> Canned beef, borax
> Canned peaches, sodium sulphite, coal-tar dye and salicylic acid
> Pickles, copperas, sodium sulphite, and salicylic acid
> Catsup, coal-tar dye and bezoic acid
> Lemon cake, alum
> Baked pork and beans, formaldehyde
> Vinegar, coal-tar dye
> Currant jelly, coal-tar dye and salicylic acid
> Cheese, coal-tar dye
> (This gives sixteen doses for supper.)
>
> ★Mark Sullivan in **Visions of Technology: A Century of Vital Debate about Machines, Systems, and the Human World**, p. 46★

The pesticides, fertilizers, and irrigation that produce high yields have environmental impacts. Crop losses to pests are enormous without some preventative measures. Organic farmers rely on non-synthetic chemicals (such as rotenone) that may be as or more toxic than synthetic compounds. Alternatives to pesticides carry their own risks. Biocontrol methods, which bring in other species to prey on crop-eating pests, have caused serious damage in some cases when the introduced species has spread rapidly in the ecosystem. Fears about genetically engineered plants have caused some countries, such as the members of the European Union, to take an extremely cautious approach to the release of genetically modified species.

In poor countries, meeting basic nutritional needs remains a serious concern. Insufficient food, which could often be solved by political changes, is a problem for many populations, but dietary shortages of important nutrients may be more serious. For example, it is estimated that almost 12 percent of the world's population suffers from iodine-deficiencies and goiters.[17] In wealthier countries, popular concern in recent years has focused on cancer risk from pesticide residue, although scientists have had difficulties establishing a firm link between pesticides and cancer. Poor diet and obesity remain the greatest food-related health risks in countries such as the United States.

Basic Concepts

Many science textbooks discuss food and agriculture issues only briefly, so it may be necessary to find supplemental resources. A college library may have textbooks in the agricultural sciences, or agronomy, which is the science of soils and crop production. Many of the concepts related to the science of agriculture, including genetics, nutrient cycling, and soils, may be covered in an upper-level biology textbook. Federal, state, and local agencies also make extensive resources available online relating to soil, agriculture, nutrition, and other related topics. We have suggested a number of sites below, in addition to books and articles that may be found in a public or university library.

Cunningham, William, and Barbara Saigo. "Food, Hunger, and Nutrition." Ch. 10 in *Environmental Science: A Global Concern*. Fifth Edition. Boston: WCB/McGraw-Hill, 1999, pp. 205-224.
In this textbook, Cunningham and Saigo summarize the advances made in agriculture in the past few decades, provide basic information on nutrients, and discuss the political and economic factors that have led to famines in the last forty years.

Arnfield, Edwin, Angela Bitterman-Wenson, and Karl Sjolander. *Environmental Science: Global Ecosystems under Stress*. Fourth Edition. Dubuque, IA: Kendall/Hunt Publishing Co., 1998, Ch. 15 & 16, pp. 261-310.
This teacher's guide discusses the history of agriculture, methods of producing and processing foods, and the challenges of growing food in different regions.

Solbrig, Otto T., and Dorothy J. Solbrig. *So Shall You Reap: Farming and Crops in Human Affairs*. Washington, DC: Island Press, 1994.
The authors provide a history of agriculture that focuses upon the scientific, social, and economic features of agricultural development as well as contemporary agricultural issues.

Diamond, Jared. *Guns, Germs, and Steel: The Fate of Human Societies*. New York: W.W. Norton & Co., 1997.
In this Pulitzer-prize winning book, noted evolutionary biologist Jared Diamond poses an answer to the question of why some societies have been able to conquer and displace others. Diamond devotes a number of chapters to the history of the development of agriculture and its effect on human societies.

Agriculture: United States

U.S. Department of Agriculture
This federal agency site includes a history of American agriculture from 1776 to 2000, and information on U.S. crops, soils, food supply and safety. *http://www.usda.gov*

Council for Agricultural Science and Technology (CAST)
This non-profit organization's mission is to communicate scientific research related to food and agriculture issues for policymakers and the media in an understandable manner. They have a number of publications online and a special section for science educators.
http://cast-science.org/scienced.htm

University of Nebraska, Institute of Agriculture and Natural Resources: Publications Index
This university site contains links to publications on agricultural topics, including nutrition, soil, and pesticides. ***http://www.ianr.unl.edu/pubs/browse.htm***

AGRICULTURE: GLOBAL

United Nations Food and Agriculture Organization: Agriculture 21
This international agency site provides publications examining the science, economics, and politics of world agriculture.
http://www.fao.org/WAICENT/FAOINFO/AGRICULT/Default.htm

World Resources Institute: Sustainable Agriculture
This non-profit research organization provides brief summaries on genetic information banks, the impact of agriculture on the environment, and food production issues.
http://www.wri.org/sustag

Jordan, T. E., and D. E. Weller. "Human Contributions to Terrestrial Nitrogen Flux." ***BioScience*** **46 (1996): 655-664.**
The authors calculate the flow of nitrogen through the land, air, oceans, and water, including contributions from human sources, particularly agriculture. Fertilizers for agriculture are a significant source of nitrogen.

AGRICULTURAL RESEARCH AND INNOVATIONS

World food production kept pace with rapid population growth in the years after World War II because of a number of agricultural innovations, especially the development of high-response varieties of food crops by agronomists such as Norman Borlaug, who received a Nobel prize for his research in this field. These innovations have been termed the "Green Revolution." Research continues to improve agricultural practices and increase yields, particularly in developing countries. Increasing yields means that more food can be grown on fewer acres, allowing water and pesticides to be used more efficiently, and alleviating the need to clear additional land for agriculture. Some "new" practices actually have long histories. Water harvesting, for example, in which land is shaped so that irrigation water runs off into collection devices, was used in the Middle East thousands of years ago. In addition, new high-tech methods are also being implemented, including using computers and sensors to monitor crops so that pesticides and water can be applied with precision.

"A New Study Finds There's Life Left in the Green Revolution." ***Science*** **261 (September 17, 1993): 1517.**
This article reports that, according to research, yields from grain crops continue to improve despite concerns that they would level off.

Crosson, Pierre R., and Norman J. Rosenberg. "Strategies for Agriculture." ***Scientific American*****, September 1989, pp. 129-135.**
Crosson and Rosenberg review the most serious environmental threats related to agriculture, including erosion, soil degradation, and salinization. They review new (and old) techniques to address these problems, but argue that institutional mechanisms, such as

property rights and incentives, are required if farmers are to conserve water, soil, and genetic diversity.

Miller, Douglass R., and Amy Y. Rossman. "Systematics, Biodiversity, and Agriculture." *BioScience* 45 (November 1995): 680-686.
The authors argue that an understanding of the interaction among organisms within agricultural ecosystems will improve attempts to control farm problems such as soil degradation and pests.

Rhoades, Robert. "The World's Food Supply at Risk." *National Geographic*, April 1991, pp. 74-105.
The author discusses the concern that the wild relatives of major food crops, which scientists use as a source of genetic material for improving crop yield, will be destroyed by expanding industrialization.

Plucknett, Donald L., and Donald L. Winkelmann. "Technology for Sustainable Agriculture." *Scientific American*, September 1995, pp. 182-186.
This article discusses how several current technologies, such as integrated pest management, genetic engineering, germ plasm banks, and computer information systems, can help the agricultural sector balance the needs of the environment with the needs of humans.

World Bank: Empowering Producer Organizations
This international organization site includes articles concerning innovations in sustainable agriculture in the tropics, such as zero tilling. Several case studies are included.
http://wbln0018.worldbank.org/essd/essd.nsf/producer/home

International Food Policy Research Institute
This organization, part of the Consultative Group on International Agricultural Research, was established in 1975 to help developing countries devise the food policies necessary to foster agricultural innovation and reduce malnutrition. The site includes a number of publications on issues related to food policies in developing countries.
http://www.cgiar.org/ifpri/index1.htm

Rosegrant, Mark, and Robert Livernash. "Growing More Food, Doing Less Damage?" *Environment*, September 1996, pp. 6-12.
This article examines how Asia might boost food production over the next twenty-five years while minimizing environmental damage.

Faeth, Paul, and Pierre Crosson. "Building the Case for Sustainable Agriculture: Policy Lessons from India, Chile, and the Philippines–Comment/Reply." *Environment*, January 1994, pp. 16-20.
Researchers at the World Resources Institute studied agricultural policies in India, Chile, and the Philippines to find practices that were both profitable and sustainable and to examine how agricultural policy affects farmers and the natural resource base.

Smith, Katherine R. "Time to 'Green' U.S. Farm Policy." *Issues in Science and Technology* 11 (Spring 1995): 71-76.
Smith argues that farm subsidies be used to encourage farmers to adopt environmentally sound practices and offers suggestions for reforming current farm subsidy programs.

Zinsmeister, Karl. "The Environmental Assault on Agriculture." ***Public Interest*****, Summer 1993, pp. 90-98.**
Zinsmeister recommends ending farm subsidies, loosening land-use restrictions, and encouraging new research in an effort to make agriculture more environmentally sound without hurting farmers.

Traditional vs. Modern Methods of Food Production

Bray, Francesca. "Agriculture for Developing Nations." ***Scientific American*****, July 1994, pp. 30-37.**
The author suggests that a smaller-scale polyculture model, may be preferable to the western model of agriculture for many developing regions. Polyculture refers to the practice of cultivating several different crop plants in close proximity to one another. Monoculture, by contrast, involves devoting large tracks of land to a single crop.

Brookfield, Harold, and Christine Padoch. "Appreciating Agrodiversity: A Look at the Dynamism and Diversity of Indigenous Farming Practices." ***Environment*****, June 1994, pp. 6-11.**
The article suggests that attempts to improve farm output in the developing world should not neglect the insights of those who use traditional farming methods, such as "slash and burn" agriculture.

Dover, Michael J., and Lee M. Talbot. "Feeding the Earth: An Agroecological Solution." ***Technology Review*** **91 (February/March 1988): 26-35.**
The authors suggest that a combination of traditional farming methods and modern biotechnology is the best method for achieving sustainable agriculture in the tropics.

Nutrition, Famine, and Hunger

In many undeveloped and developing countries, nutritional deficiencies and chronic hunger continue to be a significant problem. Sub-Saharan Africa, in particular, has been plagued with famines and hunger. It is difficult to estimate with any degree of precision the scope of poverty and hunger in the world, because, as one report says, "National governments are rarely motivated to inquire into the fate of those for whom they have not provided."[18] Despite the wide publicity given to famines in this area, their occurrence has decreased, and today famines occur when civil war or other political factors disrupt food production and distribution. Nutritional deficiencies are a more persistent concern. India, Latin America, and Asia have rapidly urbanizing populations, which cannot rely on the subsistence farming that previously sustained them. Many people in developing countries obtain as much as 90 percent of their diet from carbohydrates, obtained primarily through grains, which may lack vital nutrients. For example, rice, the staple of much of the world, provides no vitamin A. An inadequate diet increases the incidence of infections, diseases and debilitating conditions, including blindness, dwarfism, and cretinism.

Shah, Mahendra, and Maurice Strong. *Food in the 21st Century: From Science to Sustainable Agriculture*. Washington, DC: Consultative Group on International Agriculture (CGIAR), 1999.
This publication from CGIAR, a World Bank-hosted consortium of international agricultural organizations, argues that a new Green Revolution is necessary to combat hunger in coming decades. The authors point to biotechnology as the best means for increasing food production in developing countries. They recount success stories, including the insertion of a gene that produces beta carotene into the canola plant. The beta carotene relieves vitamin A deficiency, which can cause irreversible eye damage, a common problem in undeveloped countries. Available at *http://www.cgiar.org/cgrpubx.htm*

Reid, T. "Feeding the Planet." *National Geographic*, October 1998, pp. 57-74.
The article discusses food production and distribution in the United States, Africa, and China, noting that the greatest barrier to adequate nutrition is not food supply, but sociopolitical factors such as diet choices and public policies.

Von Braun. "A Policy Agenda for Famine Prevention in Africa. *Food Policy Statement*. International Food Policy Research Institute, October 1991.
This paper reports the results of a four-year study of famines in Ethiopia and Sudan. The author counters the conventional wisdom that the famines were the inevitable result of droughts, deforestation, and war. Botswana and Zimbabwe, for example, were able to avoid famines despite similar circumstances. The paper examines the accumulation of events and political factors that led to food shortages. This, and other papers on food policy, can be ordered free of charge from the Institute at *http://www.ifpri.org*

Hardin, Garrett. "Lifeboat Ethics: The Case Against Helping the Poor." In *World Hunger and Morality*. Edited by William Aiken and Hugh LaFollette. Englewood Cliffs, NJ: Prentice Hall, 1996.
Hardin, an economist known for his classic article, "The Tragedy of the Commons," uses the metaphor of a lifeboat to explain why rich nations should not provide food aid to poor nations that cannot feed their own population. At some point, he argues, you have to push them back into the water to prevent everyone from sinking.

White, Peter. "Rice: The Essential Harvest." *National Geographic*, May 1994, pp. 48-79.
This article discusses the importance of rice throughout the world, its history, methods of production and marketing, and ties to society and religion.

El-Sharkawy, Mabrouk, and Francis Mwanza. "Drought-Tolerant Cassava for Africa, Asia, and Latin America." *BioScience* 43 (July 1993): 441-451.
This article focuses on the importance of cassava as a staple food crop in Africa, Asia, and Latin America. The authors discuss the threats to it from pests and diseases, trends in cassava production, and the drought tolerance of the plant.

Ruttan, Vernon, ed. *Why Food Aid?* Baltimore: Johns Hopkins University Press, 1993.
This volume collects a number of scholarly analyses of the economics, politics, and ethics of food aid programs. The articles address the question of whether food aid is morally justified and whether it harms development of food production in recipient countries.

Africa

Ellis, William. "Africa's Sahel: The Stricken Land." ***National Geographic,*** **August 1987, pp. 140-179.**
Located along the southern edge of the Sahara desert, the land in the arid Sahel region has been degraded by overgrazing and deforestation, exacerbating the effect of drought conditions. This essay details ways in which natives are adapting their lifestyles to an increasingly barren land.

De Waal, Alex, and Rakiya Omaar. "Doing Harm by Doing Good? The International Relief Effort in Somalia." ***Current History*** **92 (May 1993): 198-202.**
The article questions the effectiveness of relief organizations in combating hunger and famine, suggesting that disaster relief is often poorly coordinated and arrives too late. This journal may be available through an electronic periodical database in a public library.

Field, John, ed. ***The Challenge of Famine.*** **West Hartford, CT: Kumanian Press, 1993.**
This edited volume examines the African famines of the early 1980s. The essays analyze the famines' causes and contributing factors and suggest steps to avoid similar catastrophes in the future.

Caputo, Robert. "Tragedy Stalks the Horn of Africa." ***National Geographic,*** **August 1993, pp. 98-120.**
The author notes that war is at the root of famine in Somalia, Ethiopia, Eritrea, and the Sudan.

China

Prosterman, Roy, Tim Hanstad, and Li Ping. "Can China Feed Itself?" ***Scientific American,*** **November 1996, pp. 90-96.**
The authors analyze the prospects for Chinese food production and survey Chinese farmers on how best to meet growing food demand.

Becker, Jasper. ***Hungry Ghosts: Mao's Secret Famine.*** **New York: The Free Press, 1996.**
Becker discusses the worst famine in all of human history, which occurred in China between 1958 and 1962. He lays the blame at the feet of Mao Zedong and the Chinese Communist Party, whose rapid collectivization program paralyzed the nation's agriculture.

Biotechnology and Genetic Engineering

Farmers have practiced a version of genetic engineering as long as they have been breeding crops and livestock. Genetic engineering, or "gene splicing," is a more direct approach that makes it possible to graft genetic material directly from one species to another, inserting into plant and animals species desirable traits enhancing pest and disease resistance, productivity, flavor, and color.

Bioengineering is seen as the hope of the future for increasing yield in developing (and developed) countries. Bioengineering resistance to pests reduces the need for pesticides and increases yield and nutrient production less expensively than other methods. Yet, despite reassurances from scientists and manufacturers that bioengineered foods are safe, there are concerns. Some ecologists, for example, raise the possibility of detrimental effects on ecosystems from species genetically engineered for heightened disease- and pest-resistance that could spread like weeds in the absence of natural predators.

U.S. Department of Agriculture: Biotechnology Information Resource
This federal government page includes bibliographies, newsletters, resource guides, information on biotech patents and policies, and educational resources.
http://www.nal.usda.gov/bic/

Access Excellence: About Biotech
This educational site, originally developed by Genentech, a bioengineering company, includes information on the history of biotech, what it is, and the legal, social, and ethical issues involved. *http://www.accessexcellence.org/AB/*

Bains, William. *Biotechnology from A to Z.* Second Edition. New York: Oxford University Press, 1998.
This reference book addresses both basic and advanced concepts in a readable and entertaining fashion.

Ronald, Pamela. "Making Rice Disease Resistant." *Scientific American*, November 1997, pp. 100-105.
This article traces the developments that led to cloning the first disease-resistant gene from rice. The author discusses the intensive effort to map the rice gene, the difficulties with introducing the genes into cells, and the final challenge of testing the transgenic plants.

Kling, James. "Could Transgenic Supercrops One Day Breed Superweeds?" *Science* 274 (October 11, 1996): 180-181.
Some biologists are afraid that the genes used to make crops resistant to viruses could escape and create a new strain of virus-resistant weeds.

Winston, Mark. *Nature Wars: People versus Pests.* Cambridge, MA: Harvard University Press, 1997.
Winston presents a balanced view of the controversy over bioengineered crops, explaining why there are concerns and the potential risks such as pest-resistance.

National Agricultural Biotechnology Council. *NABC Reports.* Ithaca, NY: National Agricultural Biotechnology Council, 1989-present.
This series, from a consortium of agricultural research institutions, includes papers that review the potential risks associated with biotechnology and assess public concerns about the technology. See *http://www.cals.cornell.edu/extension/nabc/webreport-edit.html* for a list of topics and ordering information, or call 1-607-254-4856.

FISHERIES

Fish provide a major source of protein for much of the world's population. By one estimate, fisheries account for 19 percent of total protein for humans (from animal sources) (Botsford, 509). Several important world fisheries, however, have been over-harvested. Human activities also affect fisheries indirectly by modifying their ecosystems, through agricultural runoff, erosion, and damming rivers and streams. Fisheries are a good example of what Gilbert Hardin termed, "The Tragedy of the Commons" (see Economics and the Environment), in that countries have an incentive to gather up as many fish as they can, since whatever they do not catch, another country may take. A number of treaties and conventions in the past several decades, such as the 1982 United Nations Convention on the Law of the Sea (which has not been ratified by the United States) have attempted to establish agreement on conservation measures, but it is not an easy task to monitor or enforce compliance with these agreements.

Food and Agricultural Organization. *The State of World Fisheries and Aquaculture* (SOFIA), 1998.
According to this Food and Agricultural Organization's (FAO) Fisheries Department report, fishery production in 1997 was over six times what it was in 1950. This FAO site also has extensive statistical databases on the world's fisheries.
http://www.fao.org/docrep/w9900e/w9900e00.htm

De Alessi, Michael. "Fishing for Solutions: The State of the World's Fisheries." In *Earth Report 2000*. Washington, DC: Competitive Enterprise Institute, 2000.
There is no single answer to the question of how to conserve fisheries, this author states. He discusses in some detail the incentives that have led to depletion of fish, the disputes that occur between countries such as the U.S. and Canada, and the measures some nations have taken to improve fishery management by establishing tradable quotas for catching fish. Ordering information is available at *http://www.cei.org*

World Resources Institute: Farming Fish – The Aquaculture Boom
An important source of fish for food is now supplied by aquaculture, which relieves pressure on ocean fisheries (although aquaculture can affect coastal water quality if debris from the fish ponds are released into local waters). According to this report, nearly one-third of all fish for food is produced by aquaculture.
http://www.wri.org/trends/fishfarm.html

Safina, Carl. "Where Have all the Fishes Gone?" *Issues in Science and Technology* 10 (Spring 1994): 37-43.
Safina examines the results of the Magnuson Fishery Conservation and Management Act and argues that the council empowered by the Act to manage American fisheries has not been effective.

Lawren, Bill. "Net Loss." *National Wildlife* 30 (October/November 1992): 47-52.
This article describes the economic hardship of New England fishermen facing dramatically smaller harvests and discusses the overall decline of Atlantic fisheries.

Botsford, Louis W., Juan Carlos Castilla, and Charles H. Peterson. "The Management of Fisheries and Marine Ecosystems." *Science* 277 (July 25, 1997): 509-515.
The authors argue that political factors have contributed to the decline in many important fish populations. Fisheries managers in many countries, they report, often succumb to political pressures to increase catch limits, even when physical influences such as *El Niño* have affected marine ecosystems.

United Nations: Ocean and Law of the Sea Home Page
The United Nations Convention on the Law of the Sea attempted to establish a "comprehensive regime of law and order in the world's oceans and seas"; it establishes rules governing all uses of the oceans and their resources. This page gives the text of the Law of

the Sea and its amendments, plus links to related information on marine resources, environments, and organizations. *http://www.un.org/Depts/los/index.htm*

FOOD SAFETY

Human beings practiced food decontamination and preservation even in prehistoric times, by drying, freezing, cooking, fermenting, and seasoning food. Preventing spoilage and contamination by heating foods and then sealing them in cans or jars was the result of a contest held by Napoleon Bonaparte in which an award was offered to the individual who discovered a method of preserving meat for his armies. Innovations such as pasteurization and better food handling methods, that were developed after germs were discovered by Louis Pasteur and other researchers, have reduced incidences of food-borne diseases significantly. While the food supply, especially for the richer countries of the West, has never been safer, concern over food safety is also probably greater than it has ever been before.

According to a 1990 estimate by the U.S. Public Health Service, *E. coli* O157:H7, *Salmonella, Listeria monocytogenes* and *Campylobacter jejuni* are the four most serious food-borne bacteria in the United States, based upon the severity and number of illnesses they cause. A promising but still controversial food safety method is irradiation, in which low doses of radiation are used to sterilize foods.

U.S. Food and Drug Administration: Gateway to Government Food Safety Information
This federal government page links to numerous fact sheets, studies, and news reports on food safety issues. *http://www.foodsafety.gov*

U.S. Food and Drug Administration: Estimating Exposure to Direct Food Additives and Chemical Contaminants in the Diet
This federal agency site describes how the U.S. Food and Drug Administration (FDA) assesses human exposure to contaminants in their diet. The site includes an explanation of the FDA's annual Market Basket study, which tracks trends in the levels of nutrients and contaminants in the diet of various cohorts.
http://vm.cfsan.fda.gov/~dms/opa-cg8.html

Cliver, Dean. *Eating Safely: Avoiding Food-borne Illness*. American Council on Science and Health.
This booklet describes how food-borne illnesses occur and what consumers, industry, and government can do to combat contamination. The entire booklet may be downloaded in *.pdf* format. Available at *http://www.acsh.org/publications/booklets/eating.html*

Hunter, Beatrice. "Will Beefed-up Inspection Assure Food Safety?" *Consumers' Research Magazine*, September 1993, pp. 17-21.
This article discusses how and where food contamination occurs and suggests how both individuals and industries can minimize the risks of food contamination.

National Center for Infectious Diseases: *Escherichia coli* O157:H7
This federal government site gives a detailed overview of the *E. coli* infection, including explanations on the nature of the bacterium and how it spreads, as well as symptoms, diagnosis, treatment, and prevention.
http://www.cdc.gov/ncidod/dbmd/diseaseinfo/escherichiacoli_g.htm or call the Center for Disease Control at 1-800-311-3435.

Hunter, Beatrice. "Wave of the Future? How Probiotics Fight Food-borne Illness." ***Consumers' Research Magazine*****, December 1993, pp. 19-21.**
This article reviews how probiotic (beneficial) bacteria are used to prevent and control many human ailments. The bacteria have been found to curb conditions including digestive problems, food-borne illnesses, high cholesterol levels, and colon cancer.

Avery, Denis. "The Hidden Dangers in Organic Food." ***American Outlook*****, Fall 1998, pp. 19-22.**
This article points out that people who eat organic foods are at a higher risk of certain food-borne diseases such as *E. coli* and argues that organic foods are, overall, much less safe than foods grown with synthetic pesticides. Available at
http://www.hudson.org/American_Outlook/articles_fa98/avery.htm

National Food Safety Database: Irradiation and Food
This page has a number of articles available that describe food irradiation, including genetic studies, cost analyses, and nutrition information. *http://www.foodsafety.org/bio2.htm*

Ambassade de France Ottowa: Louis Pasteur
In 1865, Emperor Napoleon III asked French scientist Louis Pasteur to investigate the diseases that were affecting French wine, causing considerable economic losses to the wine industry. Pasteur demonstrated that wine diseases are caused by microorganisms that can be killed by heating the wine to 55° C for several minutes. This process, called "pasteurization," originally used to improve the fermentation of wine and beer, was used to destroy bacteria in milk, improving its safety and reducing food-borne diseases. This site features, among other things, the contributions of French scientists, including Pasteur.
http://www.ambafrance.org/HYPERLAB/PEOPLE/_pasteur.html
An excellent site for researching the history of medicine is at
http://www.ucr.edu/h-gig/hist-science/medic.html

Tomes, Nancy. ***The Gospel of Germs: Men, Women, and the Microbe in American Life.*** **Cambridge, MA: Harvard University Press, 1998.**
It may seem remarkable now, but until the late 1800s people had no idea that invisible microorganisms were capable of causing illness. Milk and meat were regularly left at room temperature for days and people did not wash their hands before handling food. This is a fascinating history of how and when the "germ theory of disease" took hold in America.

Pest Control

Humans have sought to control both household and agricultural pests for thousands of years. The ancient Greeks, for example, used smoke to repel insects and natural poisons such as arsenic to kill them. Pesticides are, by their nature, toxic; they are intended to kill predators and pests. Without pesticides, crop losses would be severe. Some pesticides can have highly toxic or persistent effects on ecosystem or human health. Synthetic pesticides are developed as less-harmful substitutes for traditional organic pesticides such as arsenic and strychnine, but concerns about carcinogenicity of some of these substances have been raised. Plants have naturally evolved chemical defenses against predators; it has been estimated that 99.9 percent of all pesticides to which humans are exposed are from natural sources.[19] Many of these naturally-occurring plant pesticides are highly toxic and potentially carcinogenic.

University of Nebraska: Pesticide Education Resources
This university page presents a well-organized list of links to pesticide resources on the Web. *http://www.ianr.unl.edu/ianr/pat/ephome.htm*

National Research Council. *Pesticides in the Diets of Infants and Children.* Washington, DC: National Academy Press, 1993.
This study investigated whether children and infants may have more exposure or susceptibility to pesticide residues in foods and whether current regulatory practices provide adequate protection. They conclude that there is some difference in exposure due to dietary differences between children and adults. Available at
http://www.nap.edu/order/html

National Antimicrobial Information Network: General Pesticide Information
This resource links to various online publications on pesticide composition, development, use, and hazards. *http://ace.orst.edu/info/nptn/gen.htm*

U.S. Environmental Protection Agency Agriculture Center: Pesticides
This federal government site describes various types of pesticides and their ingredients, disposal procedures, and potential hazards. The site also contains links to pesticide-related laws and regulations. *http://es.epa.gov/oeca/ag/tpes.html*

American Crop Protection Association
The American Crop Protection Association site includes fact sheets and online publications on pesticides, biotechnology, and integrated pest management. The site contains descriptions of pesticide-related laws and online publications on pesticides and pesticide regulations. *http://www.acpa.org*

Wild, Alan. *Soils and the Environment.* New York: Cambridge University Press, 1993, pp. 255-262.
Wild outlines the interaction between pesticides and soils, focusing on the absorption, decomposition, and persistence of pesticides in soil.

University of California, Berkeley: The Carcinogenic Potency Project
Scientific studies on environmental hazards and cancer, including comparisons of synthetic and natural chemicals are available at this university site.
http://potency.berkeley.edu/

Gray, George M., and John D. Graham. "Regulating Pesticides." Ch. 9 in *Risk vs. Risk: Tradeoffs in Protecting Health and the Environment.* Edited by John D. Graham and Jonathan B. Wiener. Cambridge, MA: Harvard University Press, 1995, pp. 173-192.
The authors argue that EPA's current strategy of regulating one pesticide at a time is ineffective because it fails to encourage policymakers to consider whether the substitute pesticide may be more harmful than the regulated one. This policy, in effect, trades one risk for another. For example, EPA focuses disproportionately, they argue, on whether a pesticide may be carcinogenic and pays less attention to other potential health effects, such as reproductive toxicity.

Trefil, James. "How the Body Defends Itself from the Risky Business of Living." ***Smithsonian*****, December 1995, pp. 43-49.**
This article discusses the theory that toxins naturally present in our food and environment add up to quantities significantly larger than the man-made toxins we ingest through pesticide-treated foodstuffs. An abstract and related links are available at *http://www.smithsonianmag.si.edu/smithsonian/issues95/dec95/risk.html*

Pests

Revkin, Andrew. "March of the Fire Ants." ***Discover*****, March 1989, pp. 71-76.**
Red fire ants (Solenopsis invicta), known for devouring almost everything they encounter, pose a serious threat to the South, including its agriculture. The ants swarm over anything from seedlings to cattle, even people, and are becoming increasingly difficult to control due to an evolution of multiple-queen colonies.

Clark, John F. "Beetle Mania: The Colorado Beetle Scare of 1877." ***History Today*** **42 (December 1992): 5-7.**
Clark revisits the Colorado beetle scare of 1877, when the British were afraid that the beetle that decimated the Colorado potato crop would make it to their shores. Despite widespread hysteria, the beetle never conquered the island. This journal should be available through an electronic periodical database in a public library.

Genetic Resistance to Pesticides

A practical difficulty with pesticide use is genetic resistance. When a pesticide is applied, some members of the pest population may possess random genetic mutations that make them resistant to the lethal effects of the pesticide. These resistant pests will survive, breed, and pass along their resistant trait to their offspring. Eventually, a larger percentage of the pest population will contain the resistant gene. Insects evolve quickly, so after prolonged use of a pesticide, it will be less effective as the majority of the pest population develops a genetic resistance. Older pesticides, which were "broad-spectrum" (not targeted at a specific pest), made it more difficult for pests to develop genetic resistance. New pesticides are developed to attack a specific pest and, are thus, less harmful to the ecosystem, but they may allow faster development of resistance—an example of trading one risk for another.

"New Chemicals Seek to Outwit Insect Pests." ***Science*** **261 (July 30, 1993): 550-551.**
This article discusses new pesticides that may help prevent the development of pesticide resistance in insects.

Ghersa, C. et al. "Coevolution of Agroecosystems and Weed Management." ***BioScience*** **44 (February 1994): 85-94.**
This article explores how weed species change in response to the human creation of special agricultural ecosystems.

Gould, F. "The Evolutionary Potential of Crop Pests." ***American Scientist*** **79 (1991): 496-507.**
The author presents an overview of the problem of pest resistance and includes several fascinating studies of how weeds and insects evolved to cope with changes in the agricultural ecosystem.

Genetic Resistance to Pesticides
This Oregon State University course site includes an overview of genetic resistance and several interesting case studies. *http://www.orst.edu/instruction/bi301/resistan.htm*

National Research Council. *Ecologically Based Pest Management: New Solutions for a New Century.* Washington, DC: National Academy Press, 1996.
This report addresses problems with current and traditional methods of pest management, including the problem of genetic resistance. This report can be read or purchased at *http://www.nap.edu*

INTEGRATED PEST MANAGEMENT (IPM)

Concerns about environmental impacts of pesticides, and the inevitable problem of genetic resistance, have encouraged research in Integrated Pest Management (IPM), which is a combination of biological, chemical, and cultural controls. IPM can involve biological controls such as introduced predator species, alternative methods of planting that move plants away from the pests that feed on them, and genetically engineered plants. The goal of IPM is to minimize, rather than to eliminate, the use of chemical pesticides.

U.S. Department of Agriculture: National Integrated Pest Management Network
This federal government site includes links to information on pest management strategies sorted by region. *http://www.reeusda.gov/agsys/nipmn/index.htm*

Gianessi, Leonard. "The Quixotic Quest for Chemical-Free Farming." *Issues in Science and Technology 10* (Fall 1993): 29-36.
Gianessi argues that since there are few alternatives to most chemical pesticides, and non-chemical alternatives often do not completely replace their counterparts on the market, the prospects for chemical-free farming are not promising.

Stewart, Doug. "Luck Be a Ladybug." *National Wildlife* 32 (June/July 1994): 30-31.
This short article discusses the natural habits of ladybugs and how they can help gardeners and growers control such pests as the aphid and the mealybug.

Mwanza, Francis. "South American Wasp Comes to the Rescue of Cassava Growers in Africa." *BioScience* 43 (July/August 1993): 452-453.
This article describes the effort to control the mealybug and the green mite, which have decimated the cassava crop of sub-Saharan Africa. A tiny wasp introduced from South America has greatly reduced the damage from mealybugs, but the green mite continues to cause problems.

Malakoff, David. "Fighting Fire with Fire." *Science* 285 (September 7, 1999): 1841-1843.
The demand for biological control agents for pest species is increasing, this article reports, but there may be unintended consequences. The biocontrol species can become destructive invaders that pose a threat to other species in the area. This document is one of a number of articles in this issue focusing on non-native species.

Legislation

Food Quality Protection Act (Public Law 104-170)
This 1996 act amended the Federal Insecticide, Fungicide, and Rodenticide Act and the Federal Food, Drug, and Cosmetic Act. Together, the acts regulate pesticide registration, labeling, and safety standards. Available at *http://www.epa.gov/opppsps1/fqpa/*

Labs and Activities

U.S. Department of Agriculture Research Service: Experiments for the Classroom
This site presents a set of basic experiments, including testing the effects of temperature, humidity, and other factors on seeds. One exercise demonstrates the basic statistical method used in inheritance studies, the chi-square method.
http://www.epa.gov/opppsps1/fqpa/

Crop Science Society. *Experiment in Crop Science.*
This lab manual includes instructions for experiments that demonstrate applications of plant biology to crop production. Information on ordering is available at
http://www.crops.org

National Association of Biology Teachers: Shoestring Biotechnology
This program, which is still in development, is intended to utilize the "advanced techniques and sophisticated equipment of modern biotechnology to inexpensive, hands-on, investigative classroom activities." The site includes instructions for a number of "shoestring biotech" labs. *http://www.nabt.org/oldsite/index1.html*

Additional Resources

Faulkner, Edward. *Plowman's Folly.* Norman, OK: University of Oklahoma Press, 1943.
Plowman's Folly is a short seminal book about soil characteristics and agriculture methods. Faulkner was one of the first to challenge the universal farming tradition of moldboard plowing.

Sinclair, Upton. *The Jungle.* New York: Socialist Press, 1906.
In this novel Sinclair describes the unsanitary conditions of a Chicago meat-packing factory at the turn of the century. *The Jungle* is often credited as the impetus behind the first Pure Food and Drug Act.

Background

The Earth's soils are the accumulation of particles of rock, minerals, microorganisms, and organic matter, formed by the weathering of rocks and minerals below. They are rich ecosystems that support a variety of organisms and are critical in cycling nutrients. A handful of soil can contain millions of different nematodes, bacteria, protozoa, and other organisms. Soils vary in composition; some sands, clays, and silts contain little organic matter, which provide the nutrients essential for plant growth. The dark, rich soils found under grasslands and the lighter soils found under deciduous forests are generally the richest soils for farming. The Great Plains region of the central United States contains some of the richest soils for farming in the world.

....we tend to "teach nature" in unrelated segments, a collection of nice facts that look like an unworked jigsaw puzzle spread out all over the table, pieces unconnected and, for the time, meaningless. If we know the names of our neighbors and the streets in our neighborhood, why don't we also know our natural neighborhood, who and what lives there and why, who eats whom, who nurtures whom, who passes through and who remains, who's native and who's interloper?

★Ann Zwinger in **Reading the Forested Landscape: A Natural History of New England★**

What are the issues?

The quality of its soils determines whether a country or community can grow enough food to support its population, so soil is one of the most important and most vulnerable natural resources. The Romans, for example, salted the fields of their enemy Carthage to destroy its ability to feed its army. Changes in land use can have a rapid effect on soil quality. Deforestation can cause erosion, in which the layers of soil most suitable for farming are washed away. In dry areas, irrigation that is not properly managed can cause an accumulation of salts that severely degrades soil quality. Soil can be compacted by the use of heavy farming machinery, making it less permeable to water, which can exacerbate runoff and erosion. Intensive cultivation can lead to loss of soil nutrients.

What are the uncertainties?

Scientists have only begun to study the rich diversity of microorganisms that are a part of soil ecosystems, although much is understood about the vital ecosystem services that these organisms provide by breaking down organic debris (leaves, animals, and other organic materials) and by recycling nutrients. Erosion is considered to be the most serious concern, but the extent of erosion worldwide is uncertain. The United States maintains generally accurate data on soil erosion and productivity. It is known that areas in some parts of India, Asia, and Africa have had significant loss of soil, but there are disagreements about the extent of the problem.

What are the risks and tradeoffs?

Historically, civilizations have prospered or declined based on the quality of their soils. Modern farming methods, including irrigation, heavy use of machinery, and fertilizers have substantially increased the yield of agricultural products, raising nutritional levels worldwide, but they can contribute to deteriorating soil quality. Measures to conserve soil quality can be costly. Both in developing countries and the United States, farmers may not have sufficient incentives to choose long-term conservation approaches over short-term gains. Where tenure to land is uncertain, farmers are less likely to preserve the long-term health of the soil if not properly managed. When soils are exhausted, in many areas farmers clear more land, leading to increasing deforestation.

BASIC CONCEPTS

Earth science and geology textbooks may be useful for basic information about the composition of various types of soils and how they are formed; upper level biology or ecology textbooks are useful references for explanations of soil nutrients and ecosystem functions. Soil science textbooks may be available in a public or university library for more detail on these topics. However, there are extensive resources online that provide information about soil types, quality, and conservation.

Cunningham, William, and Barbara Saigo. *Environmental Science*. Fifth Edition. New York: McGraw-Hill, 1999, pp. 227-266.
The authors provide an informative chapter on world soil resources, with discussion and graphics on soil composition and soil types around the world.

Thompson, Graham, and Jonathan Turk. *Earth Science and the Environment*. Second Edition. Fort Worth, TX: Saunders College Publishing, 1999, pp. 188-209.
In this textbook, Thompson and Turk describe the basics of soil weathering and erosion, and discuss soil formation and transportation, and mass-wasting, which is the term for large downslope losses of soil such as mud slides.

Tarbuck, Edward, and Frederick Lutgens. *Earth Science*. Eighth Edition. Upper Saddle River, NJ: Prentice Hall, 1997, pp. 64-85.
This textbook covers the basics of soil structure and provides more detail than other texts in its discussion of soil types, formation controls, and mass-wasting.

Smith, Robert, and Thomas Smith. *Elements of Ecology*. Menlo Park, CA: Addison Wesley Longman, Inc., 1998
This ecology textbook provides an introduction to basic ecology concepts, including population dynamics, and nutrient and energy flows through ecosystems, including soils.

Wild, Alan. *Soils and the Environment*. New York: Cambridge University Press, 1993.
This book provides a basic introduction to the environmental chemistry of soils.

Loynachan, Thomas, et al. *Sustaining our Soils and Society.* Alexandria, VA: American Geological Institute, 1999.
This simple booklet, produced by the American Geological Institute, the Soil Science Society of America, and the U.S. Natural Resources Conservation Service, provides an excellent overview of soil science, including how soils form, the importance of healthy soils, erosion, and information on the geographic distribution of soils. Ordering information is available at *http://www.agiweb.org/pubs/*

Soil Data: United States

National Resources Conservation Service Data Resources: Soils
Originally called the Soil Erosion Service, this U.S. Department of Agriculture agency was established by Congress after the devastating destruction in the area of the Dust Bowl. Their site provides access to major national soil information databases, keys to taxonomies for classifying soil, and the geographic distribution of soil types.
http://www.ftw.nrcs.usda.gov/soils_data.html

National Soil Survey Center
This center, part of the National Resources Conservation Services, provides basic fact sheets on water quality, salinization, biological diversity, and the pH level of soils.
http://www.statlab.iastate.edu/soils/nssc/

Soil Data: International

United Nations Food and Agriculture Organization: Soil Resources Management and Conservation Service
This United Nations agency provides online access to publications and fact sheets on soils. The agency's World Reference Base for Soil Resources includes information about the classification and distribution of soil types around the world.
http://www.fao.org/ag/AGL

Department of Land Management: Soil Information
This Malaysian university site lists links to soil-related tutorial sites on soil chemistry and other soil studies. *http://www.agri.upm.edu.my/jst/soilinfo.html*

WWW Virtual Library: Soils and Substrates
This catalog site offers links to basic facts, world data, books, journals, and organizations concentrating on soil. *http://www.metla.fi/info/vlib/soils/*

Soil Ecosystems

U.S. Bureau of Land Management: The Dirt on Dirt
This agency site provides an overview of soil as an ecosystem. Brief explanations are provided of how nematodes, fungi, bacteria, and other members of soil-based biological communities live within and off different types of soils.
http://www.id.blm.gov/iso/931/soil/soil.htm

Rainforest Database, Nutrient Cycling in Rainforests: Soil and Soil Poverty
It is often thought that the thick vegetation in tropical forests means that its soils are very fertile. This is almost never the case; researchers have found that most of the nutrients in the tropical forest ecosystem are contained in the vegetation, and that frequent rainfall in the tropics washes nutrients out of soil. This site on tropical forest ecology explains the nutrient flow in tropical soils.
http://www.gn.apc.org/LivingEarth/RainforestDB/Ecology/1.8/1.8.1.html

EROSION

Erosion is a natural process caused by water and wind. It can be exacerbated by human activities such as land clearing, tilling, poor farming methods, and irrigation. In the 1930s, massive erosion caused by persistent winds, drought, and overuse of the soil resulted in huge dust storms that destroyed a large area of farmland in the South-Central United States. This disaster, partly natural and partly man-made, struck an area that became known as the Dust Bowl. One observer reported, "there was something fantastic about a dust cloud that covered 1.35 m. square miles, stood three miles high and stretched from Canada to Texas, from Montana to Ohio—a cloud so colossal it obliterated the sky..." (Goudie, 167). Practices such as terracing, conservation tillage, contour plowing, and strip cropping are now used in many places to minimize erosion and maintain soil fertility.

U.S. Department of Agriculture, Agricultural Research Service: Wind Erosion Research Unit
This federal government site includes a synopsis of what wind erosion is and why it is a problem, and additional erosion research data. *http://www.weru.ksu.edu/*

National Soil Erosion Research Laboratory: Universal Soil Loss Equation Database
This research laboratory site lists detailed storm and soil loss data from the 1930s to the present day. *http://topsoil.nserl.purdue.edu/nserlweb/usle/whatusle.htm*

Lowdermilk, Walter. "Conquest of the Land through 7,000 Years." USDA Bulletin No. 99. U.S. Department of Agriculture, Natural Resources Conservation Service, 1939.
This publication is a 1939 report written by the assistant director of the newly created Soil Erosion Service. Dr. Lowdermilk studied the record of agriculture in countries that have been under cultivation since recorded time to see what the United States could learn to avoid another Dust Bowl. He found civilizations that collapsed as a result of erosion and overgrazing, and other countries that have conserved their soils for thousands of years through careful stewardship. The publication can be ordered from the Department of Agriculture's Natural Resource Conservation Service. Information on ordering can be found at *http://www.nhq.nrcs.usda.gov/land/index/publications.html* or by calling 1-800-LANDCARE.

ADDITIONAL RESOURCES

Steinbeck, John. *Grapes of Wrath*. New York: Penguin Books, 1992 (Reissue).
This classic American novel depicts the plight of thousands of farmers in the 1930s driven off the land in the area of the Dust Bowl.

Faulkner, Edward. *Plowman's Folly*. Norman, OK: University of Oklahoma Press, 1943.
Plowman's Folly is a short seminal book about soil characteristics and agriculture methods. Faulkner was one of the first to challenge the universal farming tradition of moldboard plowing.

The Library of Congress: Voices of the Dust Bowl 1940-1941
This is an online multimedia collection of recorded interviews, photographs, manuscripts, and other publications related to Dust Bowl refugees in migrant farm camps.
http://lcweb2.loc.gov/ammem/afctshtml/tshome.html

Sears, Paul B. *Deserts on the March*. Norman, OK: University of Oklahoma Press. 1935. Washington, DC: Island Press, 1988 (Reprint).
This is an historical account of desertification. This book was widely read and influential in the 1930s.

Worster, Donald. *Dust Bowl: The Southern Plains in the 1930s*. New York: Oxford University Press, 1979.
The author provides a scholarly history of causes and effects of erosion in the Dust Bowl in the Southern Great Plains of the United States.

CHAPTER 10

Mineral Resources

Background

The comfortable existence we enjoy depends on abundant use of mineral resources. Everything we use, from pencils to computers, is made from materials that have been extracted from the Earth. The extraction, processing, and transport of minerals, though, inevitably have impacts on the environment. Mitigating disruption of landscapes and ecosystems, while ensuring supplies of critical minerals, is a technological challenge.

> It was not difficult for me to build a steam wagon or tractor. In the building of it came the idea that perhaps it might be made for road use. I felt perfectly certain the horses, considering all the bother of attending them and the expense of feeding, did not earn their keep. The obvious thing to do was to design and build a steam engine that would be light enough to run an ordinary wagon or to pull a plow. To lift farm drudgery off flesh and blood and lay it on steel and motors has been my most constant ambition. It was circumstances that took me first into the actual manufacture of motor cars. I found eventually that people were more interested in something that would travel on the road than in something that would do work on farms.
>
> ★Henry Ford as quoted in **Visions of Technology: A Century of Vital Debate about Machines, Systems, and the Human World★**

What are the issues?

Despite concerns expressed in the last few decades that some minerals resources would become scarce, supplies of most major minerals are not an immediate concern. In fact, reserves of most major minerals have increased over the last two decades despite population growth and increased consumption.

There are a number of environmental issues associated with mining and minerals extraction. Extracting minerals from the surface of the Earth or below the surface requires moving a lot of earth. Extracting one ton of copper, for example, requires moving and processing some 350 tons of ore. Although reclamation of land disturbed by mining operations is required by law, some disturbances are permanent. Processing of minerals creates a waste stream that must be carefully controlled to avoid leakages into surrounding ecosystems. In some areas, old mines abandoned before the beginning of strict regulation of mining operations pose a problem. In many cases, the companies that operated these mines are no longer in existence so it is difficult to assign liability for the costs of cleaning up the site. Access to minerals on public lands in the United States, controlled by the Mining Law of 1872, has been a continuing source of controversy.

A number of developing countries are rich in mineral resources, but lack the technology to extract the resources without degrading the environment. The oceans hold vast resources of minerals, although it is not yet cost-effective to remove them. However, several countries are racing to develop deep-sea mining technologies, and disputes over access to mineral rights in the world's oceans will undoubtedly arise.

What are the uncertainties?

The environmental impacts of mining operations are generally well understood. Research in this area focuses on the most effective methods of reclaiming and restoring lands that have been disturbed. One persistent problem is mitigating acid mine drainage, particularly from abandoned mines. Some mines dating from Roman times are still producing acid drainage.[20] Acid mine drainage comes from pyrite or iron sulfide in the ground. When disturbed by excavation, this material weathers and reacts with oxygen and water to produce high levels of iron and sulfate in runoff water. Modern mining operations use lime and other chemicals to treat the acidic drainage, but the long-term effectiveness of this method is not yet known.

What are the risks and tradeoffs?

Most products that we use today originate from mineral resources. Even most man-made materials such as plastic are derived from resources extracted from the Earth. The trend in industrial design is towards *dematerialization*, that is, using lighter weight materials, substituting man-made materials such as plastics for metals, and recycling, all of which reduce the amount of raw materials required. There are no substitutes currently available, however, for many critical materials. Gold, for example, because it is malleable, highly conductive, and does not react with other common materials, is used frequently in medical devices and is critical in computer and other electronic circuitry.

Basic Concepts

The Earth's continental crust, primarily consisting of silicate rocks, is composed of about 47 percent oxygen, with a further 27 percent being silicon. Together with aluminum, iron, calcium, sodium, and potassium, these elements make up 99 percent of the crust. Important mineral resources such as gold, lead, zinc, silver, and platinum are relatively rare; they can be mined only because certain geological processes (volcanoes, hydrothermal vents) have concentrated them in deposits of ore. These intense geological events are also associated with mountain formation; as a result many rich ores are found in mountainous states such as West Virginia, Nevada, and California.

Earth science and geology textbooks generally include explanations of the chemical composition of the Earth's crusts, and the geomorphic processes that are associated with mineral deposits. The U.S. Geological Survey provides information about the mineral resources of the United States and their location; in addition, many states have mining agencies that have information about mining operations and mineral resources in their state. In addition, the following resources may be useful.

Christianson, John W. *Global Science: Energy, Resources, Environment*. Dubuque, IA: Kendall/Hunt Publishing Co., 2000.
This textbook includes an extensive chapter on mineral resources, their formation, extraction, and use.

Schmidt, Victor, and William Harbert. *Planet Earth and the New Geoscience*. Dubuque, IA: Kendall/Hunt Publishing Co., 1998, pp. 273-293.
This textbook covers types and uses of minerals, their relative distribution in the Earth's crust and in seawater, and the geological factors and forces that contribute to formation of mineral deposits. The compact disc included with the text includes the type of satellite images used for exploration and discovery of mineral deposits.

Thompson, Graham, and Jonathan Turk. *Earth Science and the Environment*. Second Edition. Fort Worth, TX: Saunders College Publishing, 1999, pp. 21-39.
This chapter on minerals explains what a mineral is, its chemical composition, and physical properties.

Dutch, Steven, James Monroe, and Joseph Moran. *Earth Science*. Belmont, CA: West/ Wadsworth Publishing, 1998, pp 17-31.
The minerals chapter of this textbook explains the chemical composition and physical properties of minerals, the chemical composition of the Earth's crust, and makes a useful distinction between mineral reserves and resources, explaining why a gravel pit might be a better investment than a gold mine.

National Research Council. *Hardrock Mining on Federal Lands*. National Academy Press, 1999.
This report is in response to a request by Congress that the National Research Council undertake a study of hardrock mining, such as gold, silver, and copper, on federal lands. The report reviews the nature of mining, its environmental impacts, and the adequacy of current regulation of the industry. The book is available for viewing or purchase at
http://books.nap.edu

Mineral Information Institute
This industry site offers education materials most suitable for elementary and intermediate grades, but background information may be useful, including "Origin of Mineral Names" and "How Many Minerals Does It Take to Make a Lightbulb?"
http://www.mii.org/

Mineral Reserves and Resources

To a great extent, our material well-being depends on the timely recovery of the finite supplies of non-renewable minerals that exist in, on, and under the surface of the Earth. Methods of extending those supplies—i.e., making them last longer—include conservation, recycling, and substitution. But short of developing combinations of non-use, recycling, and substitution that make extraction of a given mineral unnecessary, supplies of any given non-renewable substance will ultimately be depleted.

A number of assumptions must be made to develop a timeline that will estimate how much longer supplies of any given non-renewable resource will last. It is sometimes assumed that current or recent historical use patterns will continue into the future, and that currently identified reserves constitute the total amount of any given mineral ever likely to be available for use. Such assumptions can be misleading and fail to consider more effective extraction technologies, recycling, and substitution that may affect patterns of resource use in the future.

In their efforts to assess the remaining quantities of the mineral resources of the United States and the probabilities of their availability when needed, the U.S. Geological Survey uses a classification scheme that assesses: (1) the degree of certainty about the existence and magnitude of supplies of those materials; and, (2) the economic feasibility of recovering them.

In this system, *reserves* are defined as those supplies known to exist and to be economically and technologically recoverable at the present time. Resources are defined to include all known, inferred, and theoretically possible supplies of the same substances, whether or not they are now economically and technologically recoverable. By these definitions, the resources of a given substance include its reserves, plus additional quantities that may (or may not) at some time become re-classifiable as reserves.

U.S. Geological Survey: Mineral Resources Program
This federal agency provides information regarding the mineral resources of the United States. Data include statistics on price, supply, and consumption of minerals, location of all mining operations, and remote sensing maps. *http://minerals.usgs.gov/*

Mineral Commodity Summaries
This annual report lists non-fuel mineral industry data including information on the industry, government programs, tariffs, and statistics for over 90 individual minerals and materials. Available to read or purchase at *http://minerals.usgs.gov/minerals/pubs/mcs* or call 1-888-ASK-USGS.

Kesler, Stephen E. *Mineral Resources, Economics, and the Environment*. New York: Macmillan College Publishing Co., 1994.
This college-level textbook discusses in detail the origins of mineral resources, how they are extracted, the laws that govern access to and extraction of minerals, and how mineral resources and reserves are estimated.

Hodges, Caroll. "Mineral Resources, Environmental Issues, and Land Use." *Science* 268 (June 2, 1995): 1305-1312.
Hodges reports that despite mid-twentieth century predictions by some of resource scarcity, known reserves of most major minerals have increased and improved technologies have mitigated many of the serious environmental impacts of minerals extraction in the U.S.. Developing nations, however, will need technical assistance from the U.S. if they are to exploit their mineral resources without environmental degradation.

National Mining Association. *Facts About Minerals 1999-2000*. Washington, DC: National Mining Association.
This booklet discusses mining and mineral topics including location, extraction and processing, use, and environmental concerns. The data and information are presented in a historical timeline, charts, tables, illustrations, and glossary. Available free by calling 1-202-463-2640.

Regis, Ed. "The Doomslayer." *Wired*, February 1997.
This article describes how (the now late) Julian Simon changed from a self-described anti-growth zealot to an advocate against gloomy predictions about natural resources. Included is a discussion of the famous wager that Simon made in 1980 with Paul Ehrlich, author of *The Population Bomb*. Simon, a University of Maryland professor, bet Ehrlich $10,000 that the real price of five important mineral resources of Ehrlich's choosing would be less expensive in 1990 than in 1980. Simon won the bet; all the prices fell, indicating that Ehrlich's predictions of growing scarcity within the decade were unwarranted. The *Wired* article is available online at *http://www.wired.com/wired/current.html*

Tierney, John. "Betting the Planet." ***New York Times Magazine*****, December 2, 1990, pp. 52-78.**
This 1990 *New York Times* article gave detailed coverage to the Simon-Ehrlich wager.

Minerals Extraction

Mining of copper ores began over 5000 years ago in what is today the country of Iran, although some copper tools recently found in Thailand may be even older. The world's first mining textbook, *De Res Metallica*, was written by Georgius Agricola in 1556.[21] Extraction methods vary according to the type and location of the mineral resource and available technology. Subsurface mining, or underground mining, is necessary when resources are present in deep veins. Tunnels or shafts are built to provide access. This has historically been a dangerous operation, but safety regulations and better technology have improved safety. Resources near the surface can be extracted through surface mining, or open pit mining, in which rock or soil (called overburden) is removed and resources are separated out through various methods. Environmental concerns related to this method include dust, runoff water, and groundwater contamination.

Other resources (water, oil) are extracted through drilling operations. New technologies are being developed to improve the efficiencies and mitigate the environmental impacts of extraction and processing.

American Geological Institute. ***Metal Mining and the Environment*****, 1999.**
This publication, a joint effort of the American Geological Institute and the U.S. Geological Survey, provides a brief overview of mining and how environmental impacts are mitigated, including reclamation, soil and water treatment, and remediating acid drainage. Ordering information is available at ***http://www.agiweb.org/pubs***

Enviromine
This site houses case histories, legislative updates, and explanations of various environmental technologies. The organization also maintains a long list of links to relevant sites on mining-related environmental issues. ***http://www.enviromine.com/***

Krajick, Kevin. "The Great Canadian Diamond Rush." ***Discover*****, December 1994, pp. 70-79.**
The author details the commercial and geological complexities of diamond mining in Canada's Northwest Territory. Diamonds are rare and difficult to locate, so scientists analyze physical, geological, and chemical aspects of known diamond mines to narrow the search.

Mining Environment Neutral Drainage
Implemented to develop and apply new technologies to control acidic drainage from mining operations, this organization has a long list of links to mining-related web pages, including links to mining companies and international organizations.
http://www.NRCan.gc.ca/mets/mend/

Debus, Keith. "Mining With Microbes." ***Technology Review*** **93 (August/September 1990): 50-57.**
Debus discusses ancient and contemporary methods of using microorganisms to recover and process metals. Although biohydrometallurgy is still under research, current applications have proven to be cleaner and more cost-efficient than conventional techniques.

Russell, Dick. "Deep Blues: The Lowdown on Deep-Sea Mining." ***The Amicus Journal*** **19 (Winter 1998): 25-29.**
Extensive mineral deposits of zinc, copper, silver, and other minerals exist in the deep ocean floor, which has led to a race among several nations to develop technologies to extract these minerals. This article raises a number of questions concerning deep-sea mining, including its cost, environmental impacts, and international tensions that might arise.

Watkins, T. H. "Hard Rock Legacy." ***National Geographic*****. March 2000, pp. 76-95.**
This article provides a good overview of the challenges faced by the mining industry, including falling prices, expensive cleanup operations, and rising costs, while demand for mineral resources continues to grow. The author reports that each year the U.S. consumes 24,000 pounds of new non-fuel minerals per person.

Gilbert, G.K. "Hydraulic-mining Debris in the Sierra Nevada." U.S. Geological Survey Professional Paper 105, 1917.
This is considered the classic monograph on the extensive physical impacts of hydraulic mining, which was commonly used in the Sierra Nevada in the 1850s-1880s, and elsewhere until the 1930s. This is a rather old historical document, but it can be found in public libraries that are depositories for government documents.

California Department of Conservation: Division of Mines and Geology
This office has a number of publications online and for order that tell the story of the "Gold Rush," including maps of historic gold mines.
http://www.consrv.ca.gov/dmg/index.htm

Sierra Nevada Ecosystem Study
This research project site posts the current findings of a three-year study of the ecosystems of the California Sierra Nevada Range. Several chapters address remaining evidence of the Gold Rush in the region.
http://ceres.ca.gov/snep/pubs/web/v1/v1_default.html

Dematerialization

Industrial capacity and consumption of manufactured goods increased enormously over the twentieth century; however, per capita use of raw materials, including minerals, is declining. Recycling of a number of metals is significant. For example, 66 percent of steel and 96 percent of lead from batteries are recovered and reused, which significantly reduces the need for raw materials.[22] Advances in technology including miniaturization and substitution have also reduced the amount of raw materials used in new products. Vast amounts of copper wiring used for communications networks are being replaced by fiber optic wiring made from silica, the most abundant mineral in the Earth's crust. Automobiles, home appliances and almost all consumer goods weigh considerably less than they did three decades ago, as less steel and metals are used. However, the useful lives of many of these consumer goods, such as a telephone, are less than previous designs, reversing some of the benefit over the long term.

Scarlett, Lynn. "Doing More with Less: Dematerialization—Unsung Environmental Triumph." In *Earth Report 2000: Revisiting the True State of the Planet.* Edited by Ronald Bailey. Washington, DC: McGraw-Hill, 2000.
The author reviews the trends in manufacturing over the last few decades toward lighter materials, less packaging, and more efficient use of materials. Home appliances, for example, weigh 15 to 20 percent less than they did fifteen years ago. This publication can be ordered at *http://www.cei.org/*

Allenby, Braden. *Industrial Ecology: Policy Framework and Implementation.* Upper Saddle River, NJ: Prentice Hall 1999.
This introductory college-level text introduces students to industrial ecology, which mimics natural ecology by designing manufacturing processes and products as much as possible to reduce and reuse residues of materials. The text includes a discussion of how material reductions can be made through design decisions.

Cornell University: Science News
In July 1997, Cornell University researchers announced they had made the world's smallest guitar. The guitar, which was about the size of a human cell, had six strings, although the sound would be inaudible. The nanoguitar was a project to demonstrate the potential of nanotechnology, or a technology for making tiny mechanical devices. As technology advances, materials and the size of products have been shrinking.
http://www.news.cornell.edu/science/July97/Guitar.ltb.html

Legislation and Policy

Humphries, Marc. "The 1872 Mining Law: Time for Reform?" Congressional Research Service Issue Brief for Congress, No. 89130, October 15, 1999.
This Congressional Research Service (CRS) paper describes the history of the 1872 Mining Law and proposals to amend it. Critics charge that the law is outdated; proponents counter that the proposed changes to the law would render exploration economically unfeasible. Other CRS papers on mining issues are also available at this site. Available at
http://www.cnie.org/nle/mine-1.html

Initiative to Revise and Update Regulations Governing Hardrock Mining Operations (43 CFR 3809)
This site gives the text of the Bureau of Land Management proposal to update the old 1872 regulations currently governing mining of such minerals as gold, silver, and copper on public lands. The site also tracks the legislative developments regarding the controversy. *http://www.blm.gov/nhp/Commercial/SolidMineral/3809/*

Federal Mine Safety and Health Act of 1977 (Public Law 91-173, as amended by Public Law 95-164)
Deaths and serious injuries from unsafe and unhealthful conditions were common throughout the history of the mining industry. This site gives the full text and legislative history of the act that set mandatory safety and health standards for mines.
http://www.msha.gov/REGSINFO.HTM

Mine Safety and Health Administration
This Department of Labor site has the full text of the Federal Mine Safety and Health Act, health hazard information, and mine accident, injury, illness, employment, and coal production statistics. *http://www.msha.gov*

CHAPTER 11

Population Studies

Background

It has been estimated that the global population reached 6 billion at the end of the 20th century. While historical records are full of gaps, it is thought that at the end of the 19th century, world population was about 1.6 billion, so that the number of human beings almost quadrupled in just 100 years, a scale of expansion believed to be unprecedented in human history. Moreover, most of this rapid growth occurred in the five decades following World War II.

The rapid growth in global population is due to the dramatic decline in mortality worldwide. The agricultural revolution, the availability of antibiotics, vaccines, and pesticides have transformed health—even in most developing countries—and produced an approximate doubling in human life expectancy in the past century. For example, the life expectancy of a woman born in Chile in the early 1900s was about 33 years, while the average Chilean female born today should live about 78 years.[23]

A second global transformation in global population trends is now occurring. The world population is still increasing at a rate of approximately 1.3 percent per year, adding about 78 million to the global population each year, but the global growth rate is slowing and total fertility rates are declining almost worldwide. A demographic transition is taking place. A demographic transition is the term demographers use to describe the movement from high birth rates and high death rates that typically characterize traditional societies, to the low birth rates and low death rates that are common in developed, industrialized countries.

Total fertility rates (the average number of children that a woman has over her lifetime) peaked in the period 1965-1970, when the worldwide average total fertility rate was estimated at about 5 births per woman. A replacement fertility rate would be roughly 2.1 births, or one child to replace each parent (taking into account premature deaths). For the year 2000, the global total fertility rate is projected to be 2.7 births per woman per lifetime, and fertility rates are now declining in all regions of the world. Almost half of the world's population live in countries with below-replacement fertility rates.

What are the issues?

Although the population growth rate is slowing, worldwide population is still increasing rapidly because there are large numbers of women of childbearing age in countries where fertility rates remain high. Ninety-seven percent of the annual population increase occurs in developing countries that have the lowest income levels and growing populations in these countries rely heavily on natural resources. Many countries with the most rapid population growth rates are in areas of rich biological diversity, where deforestation for firewood and cropland is particularly severe. In some urban areas, rapid population growth has outpaced improvements in facilities for water sanitation and sewage disposal, leading to water-borne and other environment-related diseases.

What are the uncertainties?

It is not a simple task to make long-term projections of population growth. Global numbers aggregate the results of personal decisions of billions of individuals worldwide. Demographers lack a robust method for making long-range estimates of population growth, although short-term projections based on current mortality and fertility rates can be made with some accuracy in the absence of any unforeseen catastrophes. Indeed, most demographers did not foresee the worldwide decline in total fertility rates that has been occurring during the last thirty years. Projecting population trends is complicated by a lack of precise data. Even in the United States, the U.S. Census Bureau estimated that the 1990 census undercounted the population by almost two percent. Many developing countries, where most of the population increase is occurring, lack the infrastructure for complete registration of birth and death statistics.[24] Demographics must rely, in many cases, on extrapolations from surveys and other sources of data.

The dynamics of changes in population trends are also not completely understood. Most modern industrialized nations have low mortality rates and low fertility rates, while poor, undeveloped countries have high mortality rates and high birth rates. In between are countries undergoing a demographic transition, in which mortality rates decline while fertility rates remain high for a period before beginning to decline. Some demographers have argued that "development is the best contraceptive," but the cause-and-effect relationships are not clear, and there are exceptions. Iran, for example, has experienced declines in fertility rates over the last two decades, even though their Gross Domestic Product (GDP) is not increasing. Education, particularly for women, is thought to be important, but countries such as Bangladesh have low literacy rates and declining fertility rates, while countries such as Egypt have high literacy rates and high fertility rates. Per capita income is correlated internationally with low fertility rates, but there are many social, cultural, and economic factors that determine trends for any particular country.

Everything has been visited, everything known, everything explained. Now pleasant estates obliterate the famous wilderness areas of the past. Plowed fields have suppressed the forests; domesticated animals have dispersed wild life. Beaches are plowed, mountains smoothed, and swamps cleaned. There are as many cities as in former years there were dwellings...Everywhere there are buildings, everywhere people, everywhere communities, everywhere life...We weigh heavily upon the world; its resources hardly suffice to support us.

★The Carthaginian writer, Tertullian, writing during the 1st century, as quoted in **Historical Ecology**★

What are the risks and tradeoffs?

Several decades ago, experts with a Malthusian[25] perspective, most notably Paul Ehrlich, predicted that the rapid population growth that took place in the years after World War II would lead to worldwide famine and starvation.[26] That has not occurred, and there has been steady improvement in health worldwide, evidenced by the decline in mortality rates and increase in average longevity in almost every country.

The worldwide increase in longevity and the decline in fertility rates have led to another demographic transition: the average age of the world population is increasing. People are living longer and fewer children are born; as a result, the percent of the total population aged 80 years or older is increasing. In Italy, for example, there are 60 percent more persons aged 65 or older than children under the age of 15.[27] This trend, which is projected to continue, means that a larger percentage of populations will rely on old-age pensions and will require medical services, an additional burden on public resources.

Although there have been major improvements in health worldwide, there are exceptions. AIDS has taken a devastating toll in sub-Saharan Africa, where average life expectancy is now declining. According to the United Nations AIDS Project, of the approximately 22 million people living with HIV/AIDS, sixty-three percent are in sub-Saharan Africa.[28]

Another serious health problem facing populations in several developing countries is malaria. During the early part of the 20th century, malaria was responsible for almost two million deaths per year, mostly in Asia, Africa, and the Pacific tropics. The discovery of DDT, a pesticide for killing mosquitoes that transmit the disease, resulted in a remarkable reduction in deaths from malaria in many countries, and it was hoped that the disease would be eradicated as a major health threat. Concerns about the safety of DDT arose, however, and international assistance for spraying DDT ceased. Malaria now accounts for about one million deaths per year, is directly responsible for one out of five deaths of children in Africa, and indirectly contributes to deaths from other diseases (World Health Organization, 49).

Basic Concepts

Extensive data on population are available on the Internet from several sources, which are noted below. The field of population studies encompasses an array of disciplines, ranging from biology and biochemistry (e.g., human fertility and its regulation) to applied mathematics, economics, sociology, and history. The basic concepts of population studies, including demographic transitions, population momentum, and doubling time are explained in simple terms in several of the resources noted below.

Cunningham, William, and Barbara Saigo. *Environmental Science*. Fifth Edition. New York: McGraw-Hill, 1999, Ch. 7, pp. 133-157.
In this textbook, Cunningham and Saigo provide a brief, clear explanation of human population dynamics, including demographic transition and total fertility rate. Also included (pp. 136-137) is an interesting discussion of how different graphs displaying the same population trends can convey entirely different impressions.

Raven, Peter, Linda Berg, and George Johnson. *Environment*. Philadelphia: Harcourt Brace Publishers, 1998, pp. 157-158.
The authors of this textbook explain formulas for calculating growth rates, doubling-time (the amount of time it takes a population to double), and the relationship between economic development and fertility.

Cohen, Joel. *How Many People Can the Earth Support?* New York: W.W. Norton & Co., 1995.
Cohen provides a clear explanation of how population trends are derived, and explains the formulas and factors for calculating future population growth and use of natural resources. He also provides an historical overview of population growth and addresses the uncertainties inherent in predicting future trends.

Historical Population Data

The rapid growth of human population is a phenomenon unique to the latter half of the twentieth century. For 2000 years the population of the world grew slowly, with periods of decline during plagues and epidemics, and periods of growth. Estimates of worldwide population figures for periods from prehistoric to modern times cannot be made with precision. There are few areas in which census information is available; historical records must be extrapolated from death records, resource use, and other historical evidence.

U.S. Bureau of the Census. "Historical Estimates of World Population."
This table summarizes the estimates that demographers have made of world population from prehistoric to modern times. Available at
http://www.census.gov/ipc/www/worldhis.html

Cohen's *How Many People Can the Earth Support?*
This book (cited above) also provides an extensive discussion of historical data on population and how the data were derived.

Livo-Bacci, Massimo. *A Concise History of World Population*. Trans. Carl Ipsen. Malden, MA: Blackwell Publishers, 1997.
This book combines historical and demographic methods, examining specific cases of human population growth and decline in great detail. It is written for a scholarly audience.

Malthus, T. R. *An Essay on the Principle of Population*. New York: Penguin Classics, 1983.
Most discussions on population growth refer to arguments made by British political economist T. Robert Malthus, whose 1798 essay on population predicted that population would increase at a faster pace than food production and famines would result. The Penguin Classics edition may be available in public libraries. The entire essay can also be downloaded as a *.pdf* file from the Electric Book Company at *http://www.elecbook.com/ebfree.htm*

Kolata, Gina Bari. *Flu: The Story of the Great Influenza Pandemic of 1918 and the Search for the Virus That Caused It*. New York: Farrar, Straus & Giroux, 1999.
Just after World War I ended, a flu epidemic struck that wiped out approximately 40 million people in less than a year and afflicted more than one of every four (and disproportionally 20 to 40 year old) Americans. The author focuses on the scientific investigation of this epidemic.

Population Statistics

Even with all the technological tools available, there is a degree of imprecision in worldwide population statistics. Accurate birth and death statistics are unavailable in many developing countries where there are unstable regimes and extensive migration of populations. Extensive data on world populations are available through the Internet. Two authoritative sources of population statistics are considered to be the United Nations Population Division and the U.S. Census Bureau.

United Nations Population Information Network
The most recent projection of population trends, *World Population Prospects: The 1998 Revisions*, is available online. *http://www.undp.org/popin/popin.htm*

U.S. Bureau of the Census
This U.S. government agency provides population data for the U.S. and worldwide. Its *World Population Profile: 1998*, a compendium and analysis of data on population, fertility, and mortality, is available online. The site also includes the International Data Base containing statistical tables of demographic and socioeconomic data for 227 countries and areas of the world. *http://www.census.gov*

Central Intelligence Agency: The World Fact Book 1999
This online document contains geographic, political, and population information for each country in the world. Printed and CD-ROM versions are available from the National Technical Information Service at 1-800-553-6847 or see
http://www.odci.gov/cia/publications/factbook/index.html

Library of Congress: Country Studies
These online books, which are available for 100 countries, provide an extensive overview of each country's history and population, including its culture, environment, literacy, and health conditions. *http://lcweb2.loc.gov/frd/cs/cshome.html*

Population Reference Bureau
Released annually, the *World Population Data Sheet* contains the latest population estimates and projections for cities and countries around the globe. The site also provides demographic data, such as fertility and mortality rates, life expectancy, and age distributions, arranged by level of development. The site includes education modules, including instructions on how to construct a population pyramid using Microsoft Excel.
http://www.prb.org

Pennsylvania State University: Population Research Institute
This university site provides links to U.S. and regional data sets for population and economic data. *http://www.pop.psu.edu/*

World Resources Institute 1998-99 Database
This CD-ROM contains statistics on population, economic, and natural resources statistics that were included in the 1996-1997 edition of *World Resources*, including 20-year time series for many variables for 170 countries, regions, and economic groupings. Includes easy-to-use software that permits user to select and export data for statistical analysis. See *http://www.wri.org* for ordering information.

Projecting Population Trends

Short-term population trends can be projected with reasonable accuracy. Under non-catastrophic conditions, it is possible to calculate, given current rates of average longevity and fertility, how many survivors and new births there will be over the short term. An important factor in the short-term growth rate of a population is its age structure, which refers to the percentages of a population at different ages. Nations with a high average age, such as many European nations, tend to face slow population growth or even population decline, because most of their population has already passed through child-bearing years. Nations with low average ages, even when fertility rates are declining, tend to face continuing population growth because a substantial fraction of their populations have yet to enter their child-bearing years. The tendency of a population to continue to grow, as a result of its age structure, even after it has lowered fertility rates, is known as population momentum. Even though fertility rates are declining almost worldwide, it can be calculated with some certainty that global population will continue to grow in the near term because of this forward momentum.

Making long-term predictions is less certain. Demographers did not predict the rapid decline in average global fertility rates that has occurred during the past three decades. Recognizing the uncertainties of long-term predictions, the United Nations Population Division offers several alternative projects: a "high," a "medium," and a "low" forecast of future population growth.

Cohen, Joel. "Ten Myths of Population." ***Discover*****, April 1996, pp. 42-44.**
The author discusses misconceptions that result from oversimplifications of the complex economic, environmental, and cultural issues involved in population projections.

Petersen, William. "Malthus: The Reactionary Reformer." ***The American Scholar*** **59 (Spring 1990): 275-282.**
Peterson analyzes the views of 18th century economist Thomas Malthus, noting that references to Malthus's work in the popular press give an incomplete picture of his thinking on human population growth.

Eberstadt, Nicholas. "World Population Implosion?" ***Public Interest*** **(Fall 1997): 3-22.**
The author examines recent United Nations population projections that the world population may begin to decline by the middle of the 21st Century and discusses the socioeconomic implications of a shrinking, aging population.

Feeney, Griffith. "Fertility Decline in East Asia." ***Science*** **266 (December 2, 1994): 1518-1526.**
Researchers can explain why fertility rates have declined in East Asia but they are not certain why the demographic changes occurred when they did. Birthrates in East Asia began to decline before the factors credited for this transition occurred.

National Research Council. ***From Birth to Death*****. Washington, DC: National Academy Press, 1999.**
This collection of scholarly articles examines the uncertainties inherent in projecting future population trends. It has long been believed, for example, that declining infant mortality rates lead to lower overall fertility rates but the mechanisms by which the two might be related are unclear and difficult to discover from available data. Available at *http://www.nap.edu*

POPULATION, POVERTY, AND THE ENVIRONMENT

A population's need for food, water, heat, and housing necessarily has an impact on natural resources. Most of the current global population increase (97 percent) is occurring in developing countries that have low per capita incomes. One theoretical model—"the vicious circle model"—illustrates the interaction between poverty, high fertility rates, and environmental degradation. For example, households in many countries rely on firewood for cooking and heat. More children can carry more firewood, but as deforestation occurs, firewood becomes scarce and children have to spend more time collecting wood. Households have an incentive to have more children, but more children collecting more firewood means more deforestation and scarcity. So much time is required for subsistence activities that there is no opportunity for education, which would be the best way to improve the children's ability to raise their income levels as adults.

There are, however, differing opinions on the extent to which it is the density of population, its levels of affluence, or some other factor that determines the relative impact a population has on the environment.

National Council for Science and the Environment: Population and Environment Linkages
This site links to articles and agency reports that examine the relationship between population and the environment. *http://www.cnie.org/pop/general.htm*

Dasgupta, Partha. "Population, Poverty, and the Local Economy." *Scientific American*, February 1995, pp. 40-45.
Although there is no *per se* causal relationship between illiteracy, poverty, resource deterioration, and high fertility rates, the variables are statistically correlated and the variables seem to have a positive feedback on one another. Dasgupta argues that fertility rates can be reduced if developing countries can improve related factors, such as literacy.

Keyfitz, Nathan. "The Growing Human Population." *Scientific American*, September 1989, pp. 119-126.
Keyfitz explains the relationship between population, agriculture, industry, and technology in developed and developing countries. He argues that access to birth control devices and information have lead to declining fertility in all countries.

Sen, Amartya. "Population: Delusion and Reality." *The New York Review of Books* 41 (September 22, 1994): 62-71.
Sen, a Nobel-prize-winning economist, examines in some detail the environmental, social, economic, and political aspects of population growth, population control policies, and immigration.

Smith, Courtland. "Assessing the Limits to Growth." *BioScience* 45 (July 1995): 478-483.
Smith proposes an equation that attempts to incorporate both optimistic and pessimistic assumptions into an assessment of the impact that population, technology, and affluence have on the environment.

Raloff, Janet. "The Human Numbers Crunch." *Science News* 149 (June 22, 1996): 396-397.
Raloff surveys the varying opinions of scientists regarding the demands placed on natural resources by growing human populations.

Moffett, George, and Mark Sagoff. "The Population Question Revisited." *The Wilson Quarterly* 18 (Summer 1994): 54-59.
This article examines answers to the question, "Is there such a thing as too many people?" proposed by current and historical figures including philosophers, economists, demographers, and biologists.

Von Braun. "A Policy Agenda for Famine Prevention in Africa." ***Food Policy Statement.*** **International Food Policy Research Institute, October 1991.**
This paper reports the results of a four-year study of famines in Ethiopia and Sudan. The author counters the conventional wisdom that the famines were the inevitable result of droughts, deforestation, and war. Botswana and Zimbabwe, for example, were able to avoid famines despite similar circumstances. The paper examines the accumulation of events and political factors that led to food shortages. This, and other papers on food policy, can be ordered free of charge from the Institute at *http://www.ifpri.org*

POPULATION POLICIES

Global population trends are the combined result of billions of individual decisions about bearing a child—a decision that is influenced by many social, cultural, and religious factors in almost as many different ways as there are different countries and ethnic groups. Policies that affect population growth, therefore, inevitably raise controversies. There is also considerable uncertainty about the dynamics of population change and which factors contribute most to declining fertility rates, so inevitably there are disagreements on policy strategies.

International Institute for Sustainable Development: Linkages
This organization has an online compilation of all the documents from the 1994 United Nations International Conference on Population and Development held in Cairo, Egypt. *http://www.iisd.ca/linkages/cairo.html*

Holloway, Marguerite. "Population Pressure." ***Scientific American*****, September 1992, pp. 32, 36, 38.**
Holloway surveys the concerns that environmentalists, women's health organizations, and human rights advocates raise concerning population policies.

Hardin, Garrett. "Lifeboat Ethics: The Case against Helping the Poor." In ***World Hunger and Morality.*** **Edited by William Aiken and Hugh LaFollette. Englewood Cliffs, NJ: Prentice Hall, 1996.**
Hardin, an economist known for his classic article, "The Tragedy of the Commons," uses the metaphor of a lifeboat to explain why rich nations should not provide food aid to poor nations that cannot feed their own population. At some point, he argues, you have to push them back into the water to prevent everyone from sinking.

Bongaarts, John. "Population Policy Options in the Developing World." ***Science*** **263 (February 11, 1994): 771-776.**
The author suggests that, in addition to family planning programs, population policies should include reducing the demand for large families by raising educational levels and the status of women. Expanded education and improved career opportunities for women could lead them to delay childbearing and limit the number of children they have.

National Council for Science and the Environment: Population Issues
This site contains Congressional Research Service Reports on topics such as immigration, U.S. population and demography, and U.S. population policy towards the developing world. *http://www.cnie.org/nle/crspop.html*

Population and Health

There has been a worldwide improvement in health and average life expectancy is increasing in most parts of the world. However, there are exceptions, particularly in sub-Saharan Africa and in some parts of the former Soviet Union. Worldwide health conditions are important both for humane reasons and for environmental impacts. In general, countries that have low mortality rates, particularly infant mortality rates, are more likely to move to lower fertility rates.

World Health Organization. "The World Health Report 1999: Making a Difference." Geneva, SW: World Health Organization, 1999.
This report, from the major international health agency, reviews the dramatic transformation in global health that has occurred over the last thirty years and details the major causes of mortality in developed and developing nations, and the major health problems that continue to be of serious concern. Information about obtaining a copy of the report, in addition to other information on world health, is available at *http://www.who.org*

Piel, Gerard. "AIDS and Population 'Control.'" *Scientific American*, February 1994, p. 124.
Piel argues that those who view the AIDS epidemic as a necessary correction to population growth are wrong. First, the epidemic will effect only a negligible change in population trends. Second, such a horrific remedy is unnecessary, as the population growth rates of even the most rapidly growing countries will eventually decline.

Eberstadt, Nicholas. "Russia: Too Sick to Matter?" *Policy Review*, June/July 1999.
This article examines the evidence of a growing health crisis in Russia. Mortality rates for men in their forties and fifties are now over three times higher in Russia than they are in Mexico. Climbing rates of injuries and cardiovascular disease could seriously exacerbate that country's economic problems. Available at
http://policyreview.com/jun99/eberstadt.html

Additional Resources

World Bank. *The Development Data Book/The Development Data Book Teaching Guide*. Washington, DC: The World Bank.
The Data Book provides demographic and economic statistics in straightforward tables, graphs and charts. The Teaching Guide consists of a series of activities and quizzes based upon material in the Data Book. Information on how to order is available at
http://www.worldbank.org

Capelle, Janice, and Marian Smith. "Using Cemetery Data to Teach Population Biology and Local History." ***American Biology Teacher*****, November/December 1998, p. 690.**
This article describes an exercise in which students collect and analyze population data from their local cemetery to connect demographic trends in their community to its local history.

Callahan, Cheryl, Alan Hoffmann, and Sherry Tipps. "Human Populations Studies or 'The Ghost of Populations Past.'" Woodrow Wilson Biology Institute, 1991.
This lab on the Access Excellence Activities Exchange site provides instructions for an exercise in which students examine and graph human population distribution and survivorship curves using data from a variety of sources. Available at
http://www.gene.com/ae/AE/AEPC/WWC/1991/human_population.html

CHAPTER 12

Waste Management

Background

Living things take in raw materials and excrete wastes that are recycled by other living organisms. Humans produce a flow of material residues that would overload the capacity of natural recycling processes, so these wastes must be managed. Solid and fluid, hazardous and nontoxic wastes are generated in our households, offices, schools, hospitals, and industries. In past centuries, people usually disposed of waste without considering health, safety, or aesthetic consequences. It was assumed, for example, that rivers and lakes would dissipate wastes and render them harmless. However, growing urban populations and the increase in the scale of manufacturing and industrial processes has made managing waste streams a significant endeavor. The Cuyahoga River in Cleveland, Ohio brought national attention to this issue when the polluted waterway caught fire in 1969. Since then, waste management has changed considerably. Federal, state, and local laws and regulations now govern every aspect of trash and waste disposal.

Every step in manufacturing – product creation, product use, and product disposal—is a more or less transient event, a temporary (possibly long-lived, but temporary) use of some set of atoms and energy... In this sense, products are just way stations in the flow of materials and energy—a temporary storage of elements and bond energy. Thus the whole sequence of events from mining or extractions to disposal is seen as a sequence of rearrangements of elements and energy.

★Robert Frosch in **The Greening of Industrial Ecosystems**★

What are the issues?

Approximately 12 billion tons of industrial wastes are generated each year, mostly (70 percent by weight) in the form of sludge.[29] Another 0.2 billion tons of municipal wastes are generated yearly (EPA, 1997). The goals of waste management are to minimize the quantity of waste, to reduce the amount of raw materials that are consumed, to dispose of non-hazardous waste as cost-effectively as possible, and to dispose of hazardous waste with minimum risk to human health and the environment. Differences arise regarding the best means of accomplishing these goals. Nearly every method of waste management can pose a threat to human health and the environment if proper precautions are not taken.

Source: U.S. Environmental Protection Agency, **Characterization of Municipal Solid Waste in the United States: 1998 Update.**

What are the uncertainties?

There are disagreements about the safety and risks associated with each waste management strategy. For example, it can be argued that waste-to-energy (burning trash to heat water or generate electricity) is an environmentally-friendly method of disposing of trash since it is a useful disposition of biomass that is commercially viable while reducing the solid waste volume by 85 to 90 percent. Some are concerned, though, about particulate matter possibly emitted by the incinerators. In addition, this method provides no incentive to reduce use of raw materials. Another area of uncertainty is the level of risks associated with former industrial sites (called brownfields), which may have contaminated soils. Redeveloping these sites for use is the subject of controversy.

What are the risks and tradeoffs?

Communications, health, safety, and convenience have been immeasurably improved by the many manufactured goods we consume. More effective design of materials and recycling can reduce the volume of raw materials required and of toxic wastes produced, but municipal and industrial waste must still be managed in ways that minimize risk to human health and the environment. Recycling reduces the amount of original resources used, but its efficiency varies according to material. In some cases, recycling consumes more energy and generates more pollutants than using virgin materials. New, technically sophisticated, environmentally sound landfills are safer, but they are costly to construct. Overall, community resistance (otherwise known as NIMBY, or "not in my backyard") has made it increasingly difficult for many cities to build new incinerators, waste-transfer stations, and landfills.

Basic Concepts

Material residues are generated from a variety of sources and a variety of forms: solid or liquid, hazardous or non-hazardous, flammable or inert. How waste is handled, and the state, local, and federal regulations that govern its management, depends on its source and characteristics. Waste from residences, commercial buildings, institutions such as hospitals and schools, and light industrial operations is collected and categorized as *municipal solid waste* (MSW). Municipal solid waste consists primarily of paper, containers and packaging, food wastes, yard trimmings, and various other inorganic wastes. Industrial waste is the waste generated from large manufacturing processes, agriculture, mining, and oil and gas production.

Waste management is a technical engineering field; there are many college-level engineering texts that provide detailed information on waste management systems, from landfills to incineration systems. For further research, these may be available at a university library. However, there are a number of readable books that discuss basic concepts related to waste management. In addition, data are readily available online for research, as noted below.

Frosch, Robert. "The Industrial Ecology of the 21st Century." *Scientific American*, September 1995, pp. 178-181.
Frosch explains how industry can imitate nature by reusing waste. Many industries are using technology to design processes and products that conserve energy and reduce waste. Ironically, current regulations require treatment and disposal of industrial waste and thus present disincentives for industries to trade or reuse residues.

U.S. Environmental Protection Agency: Municipal Solid Waste Factbook 1997
This electronic reference manual covers waste types, reduction efforts, recycling, combustion, land disposal, and related topics. Available online at
http://www.epa.gov/reg5oopa/students/municipal_solid_waste_factbook.htm

U.S. Environmental Protection Agency. "Characterization of Municipal Solid Waste in the United States: 1998 Update."
This document describes the municipal solid waste stream in the United States in 1997, and identifies trends in the types of waste generated and methods for managing waste since 1960. Available free by calling 1-800-490-9198 or at
http://www.epa.gov/epaoswer/non-hw/muncpl/msw98.htm

U.S. Environmental Protection Agency. "A Collection of Solid Waste Resources 1998."
This CD-ROM contains solid waste data and information displayed in colorful graphics. Available free by calling 1-800-490-9198.

Keep America Beautiful: Integrated Solid Waste Management
This non-profit site provides a basic introduction to waste in its online publication, "Overview of Integrated Waste Management." *http://www.kab.org/old/garbage.html*

Glenn, Jim. "The State of Garbage in America." *BioCycle* 40 (April 1999): 60-71.
This report was generated from the annual *BioCycle* survey of the fifty states and the District of Columbia on municipal waste. Detailed data on the composition of municipal waste and its management are included, in addition to an analysis of trends in waste management nationally and by state and region. Articles from this journal may be found in electronic periodical database in many public libraries.

Rathje, William, and Colleen Murphy. "Five Major Myths about Garbage and Why They Are Wrong." *Smithsonian*, July 1992, pp. 113-122.
This article reviews myths concerning landfill contents, landfill space, biodegradation, and per capita waste generation.

DeLong, James. "Of Mountains and Molehills." *The Brookings Review* 12 (Spring 1994): 34-39.
The author examines a number of issues in waste economics and dispels some common misperceptions about refuse.

Historical Background

University of the Arts: "Paris: Urban Sanitation before the 20th Century."
These University of the Arts papers are historical accounts of waste management in Paris from 1200 to 1900. *http://www.op.net/~uarts/krupa/index.html*

Ausbel, Jesse H., and Robert Herman. *Cities and Their Vital Systems: Infrastructure Past, Present, and Future*. Washington, DC: National Academy Press, 1988.
Cities have built vast infrastructures to meet the needs of their citizens. Philadelphia responded to a yellow fever epidemic by building the first municipal water system in the late 1700s, these authors report; the first modern urban sewage system was built in Brooklyn in 1857. This is an expensive publication, but it can be viewed at *http://www.nap.edu* (search by title).

Mary Gayman. "History of London Sewers" *Cleaner Magazine* (Reprinted by SwopNet Engineering Databank).
The term "sewer" meant "seaward" in Old English, and ditches were dug along city streets to drain human wastes into the Thames and ultimately to the sea. This site provides a fascinating history of London's sewers, including King Henry VIII's proclamation that every household was responsible for cleaning the sewers in front of their house, and Sir Thomas Crapper's invention of the flush toilet.
http://www.swopnet.com/engr/londonsewers/londontext1.html
An equally fascinating "History of Paris Sewers" is available at
http://www.op.net/~uarts/krupa/alltextparis.html

Strasser, Susan. *Waste and Want: A Social History of Trash*. New York: Henry Holt and Co., 1999.
Strasser describes how American attitudes about trash have changed from the 18th century, pre-industrial era to the present. Conspicuous consumption and waste were once used to display wealth, while only the poor and the thrifty recycled.

Melosi, Martin. *Garbage in the Cities: Refuse, Reform, and the Environment, 1880-1980*. College Station, TX: Texas A & M University Press, 1981.
Melosi provides an historical account of waste management in urban America.

Alexander, Judd. *In Defense of Garbage*. Westport, CT: Praeger, 1993, pp. 1-8.
Alexander discusses disposal methods, packaging, and biodegradation, as well as the economics and politics of waste management. He examines European and American waste management methods from ancient times to the present.

Reduce and Reuse

It is generally most cost-effective to reduce the quantity and toxicity of waste at its source. The trend in modern manufacturing is *dematerialization*. Products are now made with far fewer raw materials; they are lighter and stronger. Consumer products once made of steel and metals are now made of plastics and other man-made products. Compare, for example, a telephone of twenty years ago with the typical phone used today. Manufacturers reduce waste through measures such as using lighter weight materials, redesigning products to use less packaging, and substituting less toxic materials than previously used in the manufacturing stream.

Industrial ecologists seek to emulate the ecology of natural ecosystems in which waste from one organism is a source of energy and materials for another organisms. By reevaluating industrial processes and design, left over materials would not be considered "wastes," but residues that can be sold, traded, or reused. Ultimately a landfill would be regarded as a reservoir of valuable raw materials.

Resources for the Future: Solid Waste
Resources for the Future offers articles and discussion papers on the economics of source reduction, reuse, and recycling. *http://www.rff.org/environment/solid_waste.htm*

Ackerman, Frank. "Environmental Impacts of Packaging in the U.S. and Mexico." *Techné: Journal of the Society for Philosophy & Technology* 2 (Winter 1996): 1-16.
Ackerman compares the findings of two studies about the environmental effects of packaging wastes, one in the United States and one in Mexico. He concludes that, despite the seeming excess of packaging in the United States, municipal solid wastes are greater in Mexico. Available at *http://borg.lib.vt.edu/ejournals/SPT/v2_n2html/ackerman.html*

Kluger, Jeffrey. "Paper Trail." *Discover*, November 1995, pp. 54-61.
The author points out the inefficiencies of commercial food packaging and discusses how packaging has changed in response to consumer demands.

Indiana Department of Environmental Management: Indiana Materials Exchange
Because one company's waste may be raw material to another, this state exchange program maintains one list of available industrial materials and another list of materials wanted. The purpose of the exchange is to facilitate greater reuse of industrial and commercial waste. *http://www.ai.org/idem/imex/*

Allenby, Braden. *Industrial Ecology: Policy Framework and Implementation.* Upper Saddle River, NJ: Prentice Hall, 1999.
This college-level textbook provides an introduction to the principles of industrial ecology and includes case studies that illustrate how to analyze the flow of materials and energy throughout a product's lifecycle, from design to manufacturing and use and disposal.

Recycling

Recycling reduces the amount of virgin materials used in manufacturing. Many communities have recycling programs for household and office wastes, and the EPA estimates that in 1996 approximately 27 percent of the municipal solid waste in the United States was recycled or composted. Recycling efficiency varies according to the material. Some materials require more energy to recycle than to manufacture from original resources. The market for recycled materials is also volatile. When prices are too low for particular materials, it costs communities more to collect, separate, and sell them than to put them in a landfill.

Recycling in industrial uses is very effective for some materials. For example, almost every part of an automobile is recovered for use and almost 95 percent of all automobiles are recycled at least in part.[30] Over half of all lead from industrial uses is recovered. Copper, aluminum, and rubber are also recycled at high rates. Advances in design, or "design for recycling," can increase the recyclability of manufactured goods. Design considerations that promote ease of recycling include minimizing the number of different types of materials used in a product and the number of alloys (which are difficult to separate), simplifying the ways in which parts of materials are joined together (i.e., design for disassembly), and minimizing the use of toxic materials.

Strong, Debra. ***Recycling in America: A Reference Handbook.*** **Denver: ABC-CLIO, Inc., 1997, pp. 6-23.**
Strong provides a history of recycling and a discussion of recycling economics. The book has an extensive directory of print, video, and online recycling resources. Strong also discusses the EPA's hierarchy of waste management.

Ackerman, Frank. ***Why Do We Recycle? Markets, Values, and Public Policy.*** **Washington, DC: Island Press, 1996.**
Ackerman argues that recycling can rarely be justified economically, but is important for social and environmental reasons. He also discusses package design, bottle bills, and Germany's innovative "Green Dot" program.

Tierney, John. "Recycling is Garbage." ***New York Times Magazine*****, June 30, 1996, pp. 24-29, 44, 48.**
Tierney questions the emphasis on recycling as a waste management strategy and addresses common misperceptions about waste.

Grove, Noel. "Recycling." ***National Geographic*****, July 1994, pp. 92-115.**
Grove examines recycling efforts in the United States, recycling rates, and markets for recycled goods.

Keep America Beautiful. "The Role of Recycling in Integrated Solid Waste Management to the Year 2000."
This multi-volume report, prepared by Franklin and Associates, Ltd., provides data and analysis of the economics of residential and commercial recycling. The full report is available for a hefty fee, but an executive summary is offered free at
http://www.kab.org/old/garbage.html

Allenby, Braden, and Deanna Richards. ***The Greening of Industrial Ecosystems.*** **Washington, DC: National Academy Press, 1994.**
This publication from the National Academy of Engineering reviews advances in the field of industrial ecology, which studies the flows of materials and energy in industrial processes in order to minimize residues and increase efficiencies. The article, "Greening of the Telephone," makes an excellent case study of how products have been downsized. Available at *http://www.nap.edu*

Recycler's World
Recycler's World is a commercial recycling-trading site that illustrates what is bought and sold in industrial recycling. *http://www.recycle.net/*

Obvious Implementations Corporation: Internet Consumer Recycling Guide
This electronics and computer consulting firm's site provides detailed descriptions of what is recyclable and links to local recycling programs. Excerpts from *Recycling Manager*, a biweekly report, are used to illustrate average prices paid for recycled materials in the United States. *http://www.obviously.COM/recycle*

Bast, Joe. "Rethinking Recycling." ***Environmental Health Perspectives*****, November 1995.**
Bast critically evaluates the environmental and monetary tradeoffs of recycling. Available at *http://www.heartland.org/earthday96/recyclin.htm*

McCarthy, James. "Bottle Bills and Curbside Recycling: Are They Compatible?" Congressional Research Service Report for Congress, No. 93-114 ENR, January 27, 1993.
There are tradeoffs between collecting bottles from households at the curb and offering customers a deposit return if they bring bottles back. This report compares the effectiveness of curbside collection of bottles and bottle deposit programs.
http://www.cnie.org/nle/plgen-3.html

Iranpour, R. et al. "Environmental Engineering: Energy Value of Replacing Waste Disposal with Resource Recovery." ***Science*** **285 (July 30, 1999): 706-711.**
In the past, the principle concern of environmental engineers has been to dispose of wastes as safely and inexpensively as possible. These authors discuss the growing recognition that whether a substance is considered a waste or a resource depends on available knowledge and technology. They review emerging technologies to process useful substances from what have previously been considered waste streams.

Solid Waste Landfills

According to the EPA's 1997 survey of states, landfill capacity remained constant over the previous decade, and only three states report having fewer than five years' capacity for landfill disposal. Building new landfills, however, is an expensive and time-consuming process, because of both neighborhood opposition to landfill siting and federal regulations requiring sophisticated engineering to ensure safety. Over half of municipal solid waste is stored in landfills, although the overall amount is declining because of the success of efforts at recycling and source reduction. Paper products, followed by yard trimmings, make up the greatest material in landfills. For the foreseeable future, landfill storage of waste is expected to be necessary for materials remaining after recycling, reuse, and incineration.

University of Wisconsin, Madison: Solid and Hazardous Waste Education Center
This university site gives a general overview of landfill issues and provides illustrations of facility structures. *http://www.engr.wisc.edu/shwec*

Rathje, William. "Once and Future Landfills." ***National Geographic*****, May 1991, pp. 116-134.**
Rathje explores topics such as the contents of landfills, biodegradability, and common misperceptions about trash.

Sachs, Jessica. "At Play on a Field of Trash." ***Discover*****, June 1997, pp. 54-59.**
This article considers the problems associated with converting closed landfills into recreation facilities and explains safeguards used in modern facilities.

U.S. Department of Energy. "Using Landfill Gas for Energy: Projects That Pay."
This federal government document explains how decomposing garbage can be converted into electricity. *http://www.eren.doe.gov/cities_counties/landfil.html*

U.S. Environmental Protection Agency: Municipal Solid Waste Landfills
This site includes information about federal government requirements for landfill locations, design criteria, and facility regulations.
http://www.epa.gov/epaoswer/non-hw/muncpl/landfill.htm

Kluger, Jeffrey. "Oh Rubbish." *Discover*, August 1994, pp. 30-35.
This article discusses how to analyze a community's lifestyle and demographics by examining the contents of its trash.

INCINERATION

Incineration at high temperatures destroys pathogens and toxic materials in waste and significantly reduces the amount of waste that must be placed in landfills. Most waste incinerators in the United States now have energy recovery systems that use the heat generated in combustion as an energy source. Opposition to incinerators has arisen because of concerns that escaping ash particles may contain trace quantities of heavy metals, dioxins, or other substances. Waste-to-energy plants have also been opposed because they reduce the incentive to recycle materials, since communities can save more money by burning trash to generate electricity than by recovering materials for sale.

Rensselaer Polytechnic Institute: Incineration Tutorial
Rensselaer advocates incineration as an efficient and environmentally safe method of waste management. This page outlines the incineration process.
http://www.eng.rpi.edu/dept/chem-eng/Biotech-Environ/incinerator.html

National Research Council. *Waste Incineration and Public Health.* Washington, DC: National Academy Press, 1999.
This report assess the current state of scientific knowledge about the safety of incinerating waste. They conclude, among other things, that the lack of data relating to emissions from incinerators leads to considerable uncertainty in resolving this issue. Available at *http://www.nap.edu*

U.S. Environmental Protection Agency: Unified Air Toxics
This federal government site contains federal air emission regulations and technical documents relating to municipal waste combustion.
http://www.epa.gov/ttn/uatw/129/mwc/rimwc.html

Schneider, Keith. "Burning Trash for Energy: Is It an Endangered Industry?" *New York Times*, October 11, 1994, p. A18.
This article provides an overview of the environmental and monetary costs of municipal waste incineration.

Pearce, Fred. "Burn Me." *New Scientist* 156 (November 22, 1997): 30-34.
Pearce analyzes the comparative costs and benefits and suggests that incinerating paper is better for the environment than recycling it.

HAZARDOUS WASTE

Hazardous waste is defined as waste material that is flammable, corrosive, reactive, or toxic. The handling, treatment and storage of hazardous wastes is regulated under the federal Resource Conservation and Recovery Act (RCRA). Businesses that generate hazardous wastes are required by legislation to manage them from generation to disposal. Hazardous waste is treated in ways that change its biological, chemical, or physical characteristics either to make it less hazardous or to reduce its volume. Many industries are reducing hazardous waste by altering the manufacturing process or replacing hazardous materials with less hazardous or non-hazardous substitutes.

In 1980, Congress passed the Comprehensive Environmental Response, Compensation, and Liability Act (CERCLA), commonly known as Superfund. Superfund seeks to provide for the cleanup of abandoned hazardous waste sites by holding liable not only those who generated the wastes but everyone ever associated with hauling the wastes or managing the site. This has slowed the process of cleanup and fueled extensive and costly litigation. Efforts to reauthorize the Superfund program have not been successful.

The Resource Conservation and Recovery Act (42 USC sec. 6901)
Focusing on active and future facilities, this act regulates hazardous wastes from "cradle to grave." *http://www4.law.cornell.edu/uscode/42/ch82.html*

Comprehensive Environmental Response, Compensation, and Liability Act and the Superfund Amendments and Reauthorization Act (42 USC sec. 9601)
Commonly known as Superfund, these acts and amendments established regulations and requirements concerning the release or threatened release of hazardous substances that could endanger public health or the environment.
http://www4.law.cornell.edu/uscode/42/ch103.html

U.S. Environmental Protection Agency: Hazardous Waste Data
This federal government site provides information about the generation, management, and final disposal of hazardous wastes in the United States.
http://www.epa.gov/epaoswer/hazwaste/data

Agency for Toxic Substances and Disease Registry: HazDat Database
This site provides scientific and administrative information about hazardous substances at Superfund sites in the United States. *http://atsdr1.atsdr.cdc.gov:8080/hazdat.html*

Resources for the Future: Hazardous Waste and Superfund
Resources for the Future includes on its site a number of papers discussing the science, economics, and legislation of hazardous waste in the United States.
http://www.rff.org/environment/haz_waste_superfund.htm

Gibbs, Wayt. "Ounce of Prevention." Scientific American, November 1994, pp. 103-105.
Gibbs reports the success of hazardous waste minimization research funded by the National Science Foundation and the Environmental Protection Agency. He argues that industries can reduce overhead costs and minimize hazardous waste production by redesigning processes and substituting materials.

Rabe, Barry. *Beyond NIMBY: Hazardous Waste Siting in Canada and the United States*. Washington, DC: Brookings Institution, 1994.
Rabe identifies the problems associated with hazardous waste disposal and presents strategies for selecting sites. Available for purchase at
http://www.brookings.edu/press/press_hp.htm or by calling 1-800-275-1447.

Stix, Gary. "Clean Definitions." *Scientific American*, December 1993, pp. 26-28.
The author discusses the liability provision of the Superfund law, presenting the viewpoints of both environmental and industry groups.

National Science Teachers Association. *Teach with Databases: Toxics Release Inventory.* Arlington, VA: NSTA, 1998.
This program, jointly developed by the National Science Teachers Association and the U.S. Environmental Protection Agency (EPA), uses the EPA's Toxics Release Inventory (TRI) as the focus for a series of activities. The teacher's guide uses this database (which is included on a CD-ROM) for activities such as using the Law of Conservation of Mass to determine the amount of chemicals released in the environment and performing a risk analysis based on local toxic releases. For ordering information call 1-800-722-NSTA.

Chollar, Susan. "The Poison Eaters." *Discover*, April 1990, pp. 76-78.
Pollutant-eating microbes have been used at several sites to clean up hazardous wastes. The author reviews the advantages and disadvantages of this method, called *bioremediation*, which takes more time than other approaches but can be less costly than other mitigation measures.

Canby, Thomas Y. "Bacteria: Teaching Old Bugs New Tricks." *National Geographic*, August 1993, pp. 36-61.
The author describes a process called *bioventing*, in which tubes are placed in the soil to circulate air into groundwater, permitting bacteria present in the soil to break down toxins in the groundwater. He reviews concerns that introducing bacteria for bioremediation could alter ecosystems, as has occurred with other introduced species.

National Research Council. *In Situ Bioremediation: When Does It Work?* Washington, DC: National Academy Press, 1993.
This report assesses the state of knowledge about bioremediation and its potential for cleaning up hazardous wastes. The report includes a clear explanation of this method and how it has been used. Available at *http://www.nap.edu*

Legislation and Policy

Pollution Prevention Act of 1990 (42 USC sec. 13101 and 13102)
This law sets priorities for managing wastes that are known as the EPA's "hierarchy of waste management": Waste should be prevented or reduced at the source when feasible; wastes that cannot be prevented should be recycled when it is environmentally safe to do so; if wastes cannot be recycled, they should be treated to neutralize adverse affects to human health or the environment; and, if wastes cannot be managed any other way, they should be safely disposed of.
http://www4.law.cornell.edu/uscode/unframed/42/ch133.html

Resource Conservation and Recovery Act (42 USC sec. 6901)
This Act directed the Environmental Protection Agency to track movement of hazardous substances. The law also established a framework for managing non-hazardous waste.
http://www.epa.gov/region5/defs/html/rcra.htm

Comprehensive Environmental Response, Compensation, and Liability Act (42 USC sec. 9601)
More commonly know as Superfund, this law enables the Environmental Protection Agency to respond to the release, or potential release, of hazardous substances into the environment. Any party considered potentially responsible is required to contribute to

the cost of remediation.
http://www.epa.gov/superfund/whatissf/cercla.htm

Superfund Amendments and Reauthorization Act (42 USC sec. 9601)
Referred to as SARA, this law was created to reform the Superfund law enacted only six years earlier. Among other measures, SARA encourages greater state and citizen participation in the Superfund program. *http://www.epa.gov/superfund/whatissf/sara.htm*

McCarthy, James. "Solid Waste Issues in the 105th Congress." Congressional Research Service Issue Brief for Congress, No. 97006, December 23, 1998.
This report examines the federal government's role in solid waste management. McCarthy addresses the interstate shipment of waste, municipal waste trends, and related policy issues. Available at *http://www.cnie.org/nle/waste-16.html*

LABS AND ACTIVITIES

Krumhardt, Barbara, and Danielle Wirth. "Solid Wastes: Recycling." In *Experiences in Environmental Science*. East Dubuque, IL: Bellwether-Cross, 1999, pp. 137-146.
In this exercise, students are asked to determine the percentage of classroom wastes that are recyclable and to collect and analyze data from local recycling facilities.

Schug, Mark, John Morton, and Donald Wentworth. *Economics and the Environment*. New York: National Council on Economic Education, 1997, Lesson 18, pp. 124-125.
The recycling activity described in this lesson introduces students to the economic concepts of incentives and voluntary exchange. For ordering information call 1-800-338-1192.

ADDITIONAL RESOURCES

Rathje, William, and Cullen Murphy. *Rubbish! The Archaeology of Garbage*. New York: HarperCollins, 1992.
The authors explore contemporary social and economic behavior by analyzing garbage. They compare the findings of a landfill excavation with the public's perception of landfill content.

Luton, Larry. *The Politics of Garbage: A Community Perspective on Solid Waste Policy Making*. Pittsburgh: University of Pittsburgh Press, 1997.
Luton analyzes the political, cultural, historical, and ecological considerations that go into making solid waste policy in Spokane, Washington.

CHAPTER 13

Toxicology and Risk

Background

We live in an uncertain world where it is impossible to predict every potential threat, because we have imperfect knowledge. Some activities are inherently risky, such as driving too fast on crowded streets or bungee jumping, and we know that some substances, such as strychnine, are dangerous even in small quantities. But many other dangers are difficult to predict. Earthquakes and tornadoes, for example, despite years of study and improvements in technology, still strike unexpectedly and cause serious damage and suffering.

Risk can be defined as the probability and magnitude of harm from a hazard. We intuitively face and assess a number of risks each day. Perceptions of risk are inevitably subjective. Studies show that individuals and experts characterize potential dangers from risks quite differently. In one survey, for example, the public cites nuclear power as the most serious risk.[31] Health statistics, on the other hand, show that exposure to x-rays is far more dangerous. Popular perception of risk seems to be more related to the scale of potential danger (i.e., a nuclear explosion) and the ability to control the risk than to the actual statistical likelihood of suffering harm, based on quantitative evidence. Assessing and controlling risks from harmful substances in the environment is a major focus of current environmental policy.

What are the issues?

Many natural and man-made toxins and substances have the potential to harm human health and ecosystems. In the past fifty years the production of synthetic chemicals has increased significantly. Many synthetic pharmaceuticals have provided enormous benefits to human health, while synthetic materials have made easily affordable food, clothing and material goods widely available. Some synthetic substances, however, have proven to have adverse consequences at some levels of exposure, while new uses for many toxic natural substances, such as lead and potassium, have increased their impact on the environment.

All substances can be toxic at a certain quantity. Aspirin and even oxygen can be toxic at high levels. And in trace amounts, some toxic materials, for example the heavy metals chromium, copper, and zinc, are vital human nutrients. The essential question is to determine at what dose a particular substance may cause damage to a living organism. Toxic effects may be acute or occur in a short time, or may be the result of long-term or chronic exposure. Substances have widely different chemical properties that determine how they are dispersed in the environment. Some evaporate quickly and pose little harm; others are highly soluble and therefore move easily into bodies of water. Others are persistent; that is, they do not break down easily in the environment and can be harmful to bird and animal life if, after they are ingested, they leave the body very slowly.

What are the uncertainties?

Scientists cannot determine with certainty the degree of harm to an exposed person from particular substances (although the effect of some poisons are well known), but they can determine a range of likely probabilities and types of harm from different levels of chemical exposure. For example, in a report on asbestos, the National Academy of Sciences reported that for non-smoking males, the risk of lung cancer from exposure to asbestos ranges from zero additional cases per million people to 22 cases per million. There are numerous variables that contribute to the effect of a particular substance on an individual (human or animal), including the amount of exposure, exposures to other substances that reduce or increase the effect, and the general health of the individual. The risk of lung cancer from exposure to asbestos, for example, is much greater for smokers than for non-smokers.

But granting all the arguments for more industrial, biological, and medical research...one crucial question remains: Will it solve or even facilitate solutions of the problems of human relations? ...Can science save us? Yes. But we must not expect physical science to solve social problems. We cannot expect penicillin to solve the employer-employee struggle, nor can we expect better electric lamps to illuminate darkened intellects and emotions. We cannot expect atomic fission to reveal the nature of the social atom and the manner of its control. If we want results in improved human relations we must direct our research at these problems.

★George A. Lundberg, 1945, as quoted in **Visions of Technology: A Century of Vital Debate about Machines, Systems, and the Human World★**

Additional uncertainties arise from methods of estimating toxicity or other effects. It is not possible to directly test the toxicity of a substance on human populations (although some evidence is available from accidental exposures), so animal testing is generally used. These tests require giving massive doses to animals in a short period of time in order to assess levels of harm. Potential harm to humans is based on extrapolating from these massive dose levels to a level that a human would typically encounter, using mathematical models based on certain assumptions. There is some question as to whether it is the massive dose which may be harmful in itself, and that the degree of risk from small doses of a substance may be overstated. This is a difficult question, because human exposure is often thousands or even tens of thousands of times lower than doses given to tested animals. Moreover, the differences between animal and human physiology may be significant; thalidomide, for example, caused no adverse effects on animals tested, but caused severe birth defects in humans.[32]

What are the risks and tradeoffs?

In the past few decades, regulatory policy has tended to focus on the potential carcinogenic effect of chemicals, while paying much less attention to other potential effects on human health or on ecosystems. It is also implicitly assumed that synthetic materials are less safe than natural materials, which is not necessarily the case. Aflatoxin, for example, a naturally occurring mold that grows on peanuts and corn, is highly carcinogenic.

Tradeoffs must be made in managing and reducing risks. Many cancer drugs, such as Tamoxifen, are themselves carcinogenic. DDT saved millions of lives during and after World War II by killing the mosquitoes that carry malaria, but is no longer used in many countries because of concerns about its effect on bird life. As a result, malaria has reemerged as a serious health problem in many countries, causing millions of deaths. Rational choices about managing risks require scientific information, but are inevitably subject to ethical, political, and social considerations.

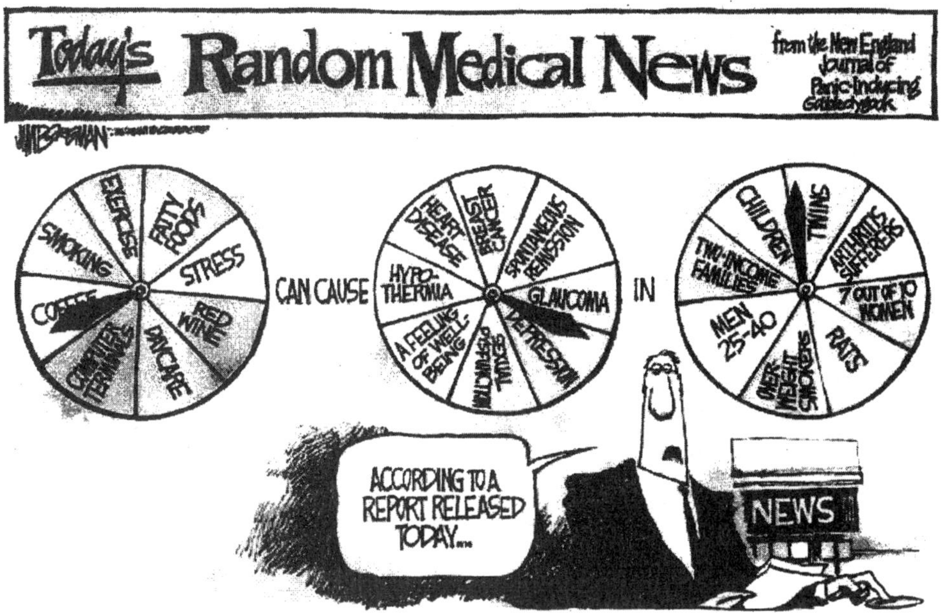

Basic Concepts

Risk assessment is the process of evaluating potential harms from substances, activities, or natural phenomena, so that they can be reduced, avoided, and rationally managed. To evaluate the potential of adverse consequences from a particular chemical or other substance, scientists have to assess the toxicity of the substance, that is, determine if and in what amount a substance may cause harmful effects to humans and to ecosystems. They also have to evaluate the degree to which people are exposed to the substance. The toxicity of a substance may be assessed in various ways, including studies of its effect on animals, studies of its effect on human cells, and epidemiological studies. Exposure assessments estimate how much and how people may have been exposed to a substance, considering factors such as how quickly a substance disperses or breaks down, and whether the substance is inhaled, ingested, or absorbed through the skin. There are many uncertainties in these estimates, so risk assessments should always be stated as a range of probabilities.

There are several excellent primers available online that explain basic concepts related to risk assessment, toxicology, and other related topics. There are also several journals devoted to risk and toxicology issues, many of which are available online. In addition there are several books noted below which explain the basic concepts in clear, understandable terms. Risk assessment is a quantitative discipline, which distills scientific information so that it is useful for policy making. Several of the resources noted below provide an introduction to significance levels, cause and effect relationships, probability, and other key concepts, but it would also be useful to refer to one of the books that provide a basic introduction to statistics.

Reporting on Risk: A Journalist's Handbook on Environmental Risk Assessment
This useful primer explains basic risk concepts, including how toxicity is assessed and how to test the validity of a study.
http://www.facsnet.org/report_tools/guides_primers/risk/

Ottobono, M. Alice. ***The Dose Makes the Poison: A Plain-Language Guide to Toxicology.*** **Second Edition. New York: van Nostrand Reinhold, 1997.**
Ottobono, staff toxicologist with the California State Department of Health, explains in very clear terms the factors that make chemicals harmful or harmless. She explains, among other things, the dose-response relationship and how factors such as age, gender, and type of exposure influence toxicity.

Wilson, Richard, and E. Crouch. "Risk Assessment and Comparisons: An Introduction." ***Science*** **236 (April 17, 1987): 267-270.**
The authors provide a basic introduction to risk assessment, including how risks are estimated and compared. Perceptions of risk, they demonstrate, can vary according to how the risks are communicated.

Risk Assessment: A Flexible Approach to Problem Solving. **Arlington, VA: Chemical Manufacturers Association.**
This 26-page booklet explains what risk assessment is in simple terms, covering hazard identification, dose-response assessment, exposure assessment, and risk characterization. For ordering information call 1-800-262-8200.

Hahn, Robert, ed. ***Risk, Costs, and Lives Saved: Getting Better Results from Regulation.*** **New York: Oxford University Press, 1996.**
This collection of articles discusses the various scientific, economic, and policy issues underlying risk assessment and management.

Brody, Jane. "Personal Health: A Study Guide to Scientific Studies." ***New York Times,*** **August 11, 1998.**
This article explains the differences between laboratory experiments, epidemiological studies, and clinical trials, to help the reader become a more discerning judge of health information presented in the media. Available at
http://www.junkscience.com/news3/brody.htm

Marczewski, Alice, and Michael Kamrin. ***Toxicology for the Citizen.*** **Institute for Environmental Toxicology, Michigan State University, 1991.**
This short booklet, intended for the general public, describes simply what toxins are, how they enter the body, and how toxicity is determined, including brief discussions on routes of exposure, dose-response curves, and establishing standards. Available at
http://www.iet.msu.edu/

Blumberg, Avrom A. "Risks and Chemical Substances." ***Journal of Chemical Education,*** **November 1994, pp. 912-918.**
This is a clear explanation of important concepts related to risk assessment, including epidemiological methods, the limitations of animal testing, and the distinction between toxins, toxicity, and probabilities of risk.

Phillips, John. ***How to Think About Statistics.*** **New York: W.H. Freeman and Co., 1999.**
In newspaper articles, reports, and textbooks, statistics are often used to demonstrate a point and a careful reader needs to understand how the numbers are derived. In this book, Phillips explains basic statistical concepts step by step to show how numbers are gathered, interpreted, and tested. He also provides examples of how statistics can be manipulated.

Slavin, Steve. ***Chances Are: The Only Statistics Book You'll Ever Need.*** **New York: Madison Books, 1998.**
This accessible book provides an easy to understand introduction to statistics.

Perceptions of Risk

Slovic, Paul. "Perception of Risk." ***Science*** **236 (April 17, 1987): 280-285.**
Slovic discusses the disparities between actual risks, perceptions of risk, and acceptance of risks. People generally accept a high degree of risk from acts that are voluntary such as skiing. They are less tolerant of involuntary risks such as those associated with food preservatives.

Singer, Eleanor, and Phyllis Endreny. "Reporting on Risk: How the Mass Media Portrays Accidents, Diseases, and Other Hazards." ***Risk*** **5 (Summer 1994).**
The authors compare findings of studies conducted in 1960 and 1984 that examine media coverage of risk. They found that most stories covered a dramatic recent event rather than cumulative risks and that much of the coverage contained inaccuracies, omissions, or oversimplifications. Available at
http://www.fplc.edu/risk/vol5/summer/singer.htm

Friedman, Sharon. "The Media, Risk Assessment, and Numbers: They Don't Add Up." ***Risk*** **5 (Summer 1994).**
According to Friedman, the complex nature of risk assessment accounts for the misleading coverage of risk numbers in the media. When actual or implied numbers are used, reporters rarely provide sufficient information for the public to understand the significance of the issue. Available at *http://www.fplc.edu/risk/vol5/summer/friedman.htm*

Thompson, Kimberly. "Health Insight: A Consumer's Guide to Taking Charge of Health Information." ***Risk in Perspective*** **7 (October 1999): 1-6.**
This publication, from the Harvard Center for Risk Analysis, gives the reader ten questions to ask of the health information they regularly obtain through various media sources. The article suggests the reader scrutinize both the source and accuracy of the data, as well as the resulting tradeoffs of the reported information.
http://www.hsph.harvard.edu/insight/guide.html

Information on Risk-Related Issues

Franklin Pierce Law Center: Risk—Health Safety & Environment.
This site provides full-text articles from *Risk*, the journal of the Risk Assessment and Policy Association, covering a wide variety of risk topics.
http://www.fplc.edu/tfield/rskindx.htm

RiskWorld
This Society for Risk Analysis site provides full-text access to articles and reports on issues related to risk. *http://www.riskworld.com/Default.html*

Harvard Center for Risk Analysis
The Center provides access to several of its articles and publications on risk-related topics. *http://www.hsph.harvard.edu/Organizations/hcra/hcra.html*

National Institute of Environmental Health Sciences
This institute, part of the National Institute of Health, focuses on environment-related diseases and risks. The Institute site includes fact sheets and questions and answers about environmental health topics such as asbestos, electromagnetic fields, dioxin, aflatoxin, asthma, and cancer. *http://www.niehs.nih.gov/*

American Council for Science and Health. *Priorities For Health.*
This online periodical by the American Council for Science and Health includes articles by scientists and physicians that address health concerns that are often misreported in the media. Recent issues have included articles on environmental health issues such as biotechnology, electromagnetic fields, multiple chemical sensitivity, and ethnobiology.
http://www.prioritiesforhealth.com

Jaworowski, Zbigniew. "Radiation Risk and Ethics." *Physics Today*, September 1999, pp. 24-29.
In this provocative article, the author argues that the current assumption that radiation at any level is dangerous has led to irrational fear and has imposed enormous costs, reducing the amount available for other life-saving measures. Available at
http://www.riskworld.com/Nreports/1999/jaworowski/NR99aa01.htm

Graham, John, and Jonathan Wiener, eds. *Risk vs. Risk: Tradeoffs in Protecting Health and the Environment*. Cambridge, MA: Harvard University Press, 1997.
The editors have compiled nine case studies illustrating the dilemmas posed by efforts to reduce risk. Policies designed to reduce risk sometimes result in transferring risk from one population to another. Recycling lead batteries, for example, is very successful and reduces environmental impacts from mining and processing new supplies of lead. It, however, increases exposure to lead emissions for those who live near lead-recycling facilities.

Ott, Wayne, and John Roberts. "Everyday Exposure to Toxic Pollutants." *Scientific American*, February 1998, pp. 86-91.
Environmental regulations target industrial and automobile sources of pollutants because they seek to control emissions of potentially dangerous pollutants, not the actual exposure that humans may have to them. Studies indicate, however, that people are not at the greatest risk outside, but indoors in places like their home and place of employment. One study revealed, for example, that industry sources were only responsible for about 3 percent of human exposure to benzene. Available at
http://www.sciam.com/1998/0298issue/0298ott.html

Gray, George M. "Measuring Risk, not Just Emissions." *Regulation* 22 (Fall 1999): 12-15.
The Toxic Release Inventory (TRI), established under the Emergency Planning and Community Right-to-Know Act of 1986, is intended to indicate environmental performance by companies. Gray argues, however, that the TRI can be misleading because it reports only pounds of emissions released rather than the actual risk those emissions may pose. Current models assess the effects of a chemical on a hypothetical Maximally Exposed Individual, which may overstate risk by 100 to 1000 times. He argues that reporting would be more useful if it including only emissions known to pose a significant risk. Available at *http://www.cato.org/pubs/regulation/regv22n4/gray.pdf*

Lieberman, A., and Simona Kwon. "Facts versus Fears: A Review of the Greatest Unfounded Health Scares of Recent Times." American Council on Science and Health.
This online report summarizes some of the most publicized recent health scares, such as red dye No. 2, which prompted Mars candy company to stop manufacturing red M&M's, even though they did not contain the dye. The authors review the charges, assess the evidence, and report what the most current and credible scientific studies indicate. Available at *http://www.acsh.org/publications/reports/factsfears.html*

Roberts, Leslie. "Dioxin Risks Revisited." ***Science*** **251 (February 8, 1991): 624-626.**
Based on their understanding of how dioxin binds to and activates receptors, a group of scientists are challenging the EPA's linear model for assessing the risks of dioxin. This article describes how the scientists arrived at their decision, the differing views that still exist among them, and the implications for regulatory policy.

Schierow, Linda-Jo. "Dioxin: Reassessing the Risk." Congressional Research Service Report for Congress, No. 95-1059 ENR, 1998.
This briefing for Congress reports on the Environmental Protection Agency's reassessment of the risks related to dioxin, based on studies that indicate animals may be more sensitive to dioxin than humans. *http://www.cnie.org/nle/pest-7.html* Related EPA documents can be found at *http://www.epa/gov/nceawww1/dioxin.html*

Berry, Richard M., Corinne E. Luthe, and Donald H. Voss. "Ubiquitous Nature of Dioxins: A Comparison of the Dioxins Content of Common Everyday Materials with that of Pulps and Papers." ***Environmental Science and Technology*** **27 (1993): 1164-68.**
The authors find that dioxins are prevalent in the environment, suggesting multiple sources, human and natural.

Goldberg, Alan M., and John M. Frazier. "Alternatives to Animal Testing." ***Scientific American*****, August 1989, pp. 24-30.**
This article describes the traditional methods for assessing the toxicity of a substance. Concerns about the limitations of animal testing, which is a lengthy and expensive process, have led researchers to consider other methods, such as *in vitro* testing, in which researchers study the effects of a substance on living cells. This method, the authors point out, also has limitations. Toxicologists cannot fully anticipate all of the interactions that might take place in a human body from evidence based on *in vitro* tests.

National Agricultural Biotechnology Council. ***NABC Reports*****. Ithaca, NY: National Agricultural Biotechnology Council, 1989-present.**
This series, from the conference proceedings of a consortium of agricultural research institutions, includes papers that review the potential risks associated with biotechnology and assess public perceptions of this technology. See
http://www.cals.cornell.edu/extension/nabc/webreport-edit.html for a list of topics and ordering information or call 1-607-254-4856.

Hafenmeister, D. ***Biological Effects of Low-Frequency Electrical Fields*****. College Park, MD: American Association of Physics Teachers, 1999.**
This is a collection of twenty articles from scientific journals that investigate possible health effects from electromagnetic fields. Available for purchase at *http://www.aapt.org*

EPIDEMIOLOGY

Epidemiology is the study of patterns of disease in populations. Epidemiologists compare the similarities and characteristics of a group that has a disease with other groups that do not, in order to investigate the characteristics that might be associated with the disease. Epidemiology has been used successfully to investigate sources of some diseases, including cholera and Legionnaires' disease (see Indoor Air Quality), and can help identify risk factors associated with diseases. Epidemiological studies provide only evidence, however. They rarely provide proof of a causal relationship between exposure and disease. Uncertainties arise in epidemiological studies in a number of ways. The population studied may be too small or biased in some way. For example, if rates of heart disease in a group of construction workers are compared to rates of the disease in the general population, they may be lower if construction workers tend to get more exercise than the population at large. Individuals in the population may not remember details that are required for comparisons of characteristics of two populations, such as their eating habits or health symptoms from years past. An epidemiological study is an important first step in assessing risks, but cannot establish what causes a disease or an effect.

British Medical Journal: Epidemiology for the Uninitiated
This is an excellent overview of the science of epidemiology for the layperson.
http://www.bmj.com/epidem/epid.html

Foundation for American Communications: Epidemiology for Journalists
This site for journalists has an online tutorial explaining the basic concepts of epidemiology. It explains how studies are conducted and includes an easy-to-understand overview of statistics and analytical methods, and a glossary of epidemiological terms.
http://www.facsnet.org/report_tools/guides_primers/epidemiology/

UCLA School of Public Health, Department of Epidemiology
This university department has compiled an extensive site on the history of epidemiology, including information on the first epidemiologist, Dr. John Snow, who identified the source of the cholera plague that began in London in 1848. His original scientific paper on the transmission of cholera is available here. *http://www.ph.ucla.edu/epi/snow.html*

Supercourse: Epidemiology, the Internet, and Global Health
This site is designed to provide an overview on epidemiology for medical and other students around the world. Over 1200 academic faculty members from 101 countries have posted their lectures here for global access. *http://www.pitt.edu/~super1/*

Palfreman, Jon. "Apocalypse Not." ***Technology Review*** **99 (April 1996): 24-35.**
Using electromagnetic fields as an example, Palfreman discusses the complexity of establishing a causal relationship between a substance and a disease. The "Hill Criteria" maintains that causation should only be inferred when there is a biologically plausible link to an association.

CARCINOGENS

Cancer is a dreaded disease, for good reason, and concerns have been raised about the potential carcinogenic effect of many substances in the environment. Reported cancer incidence rates in the United States increased in the years following World War II for a number of reasons. Cancer is particularly a disease of old age and became a significant cause of mortality in the United States when more people began living long enough to die from cancer rather than dying at earlier ages in childbirth or from tuberculosis or other diseases. Reported cancer rates also increased because of significantly improved diagnostic capabilities in the past few decades. In earlier decades, the cause of death for many individuals was not known or attributed to other causes, rather than to cancer. However, cancer mortality has declined in recent years for almost all cancers.

There are a number of potential carcinogens, both natural and synthetic, in the environment. Testing for carcinogenicity is usually done using lab rodents, which are exposed to massive doses of a substance for their entire two-year lifetime. Questions have been raised about whether the massive dosing of animals may itself have carcinogenic effects, and may lead to an overestimate of risks from lower levels of exposure.

Centers for Disease Control and Prevention
This government agency tracks the incidence of disease and mortality in the United States. The site includes useful risk-related materials including their *Mortality and Morbidity Weekly Report*, Behavioral Risk Factor Surveillance System, and the Hazardous Substance Release/Health Effects Database. *http://www.cdc.gov/*

National Cancer Institute
The National Institute of Health's site provides access to information on cancer risk and prevalence, including information on cancer clusters. *http://www.nci.nih.gov/*

National Research Council. *Carcinogens and Anticarcinogens in the Human Diet: A Comparison of Naturally Occurring and Synthetic Substances*. Washington, DC: National Academy Press, 1996.
This report outlines the results of the NRC's review of the available scientific evidence concerning the comparative cancer risk from naturally occurring and synthetic substances, such as pesticides. The report concludes that human exposure is potentially far greater to naturally occurring pesticides and other potential carcinogens than to synthetic ones, particularly since synthetic substances are regulated. They also note that the benefit from a diet rich in fruits and vegetables may counterbalance potential risks from pesticide residue on produce. To read or purchase see *http://www.nap.edu*

Caldwell, Mark. "Beyond the Lab Rat." *Discover*, May 1996, p. 70-75.
This article explains the limitations of using rodents to test the toxicity of a substance and discusses molecular epidemiology, a developing scientific field that studies the biological mechanisms by which a suspected substance affects human cells.

Gold, Lois et al. "Rodent Carcinogens: Setting Priorities." *Science* 258 (October 9, 1992): 261-265.
The authors question the utility of using animal tests to determine the risk substances pose to humans. They argue that environmental policy tends to discount the fact that humans are exposed to more natural toxins than synthetic, although the toxicology of natural and synthetic toxins is fundamentally similar.

Ames, Bruce, Renae Magaw, and Lois Gold. "Ranking Possible Carcinogenic Hazards." *Science* 236 (April 17, 1987): 271-277.
Although rodent cancer tests are not always accurate predictors of human risk, the tests are not without utility. The authors designed research models to account for differences between dose and size in humans and lab rats and to permit ranking of the potential hazard from known carcinogenic substances.

Environmental Health Information Services
This site is a service of the National Institute of Health and the National Toxicology Program. The full text of the most recent *Report on Carcinogens*, which lists every known carcinogen, is available here, in addition to other issues on environmental health topics. *http://ehis.niehs.nih.gov/*

National Toxicology Program
This federal program, part of the Department of Health and Human Services, was established to coordinate federal toxicology research programs and to provide information to federal agencies and the public about potentially toxic materials. Reports on research on carcinogens, animal testing, and emerging issues such as phthalates and endrocrine disrupters are available online. *http://ntp-server.niehs.nih.gov*

Agency for Toxic Substances and Disease Registry
This federal program, part of the Centers for Disease Control, provides toxicological profiles of substances identified as potentially harmful, including a ranking (as required by federal legislation) of the twenty most hazardous substances.
http://www.atsdr.cdc.gov/

Gawande, Atul. "The Cancer-Cluster Myth." *The New Yorker*, February 8, 1999, pp. 34-37.
When a community is afflicted with an unusually high number of cancers, the author reports, its citizens start looking for a reason—in the ground, the water, or the air. This lengthy article reviews several "cancer clusters" that have been reported by the media and describes how scientists differentiate true clusters from unfortunate random occurrences. The National Cancer Institute also has site on cancer clusters that explains the issues involved (see above). *http://www.graylab.ac.uk/cancernet/600358.html*

Bettis, Clifford and Carl Throckmorton. "What Teachers Should Know about Radon." *The Physics Teacher*, September 1991, pp. 338-343.
The authors give an introduction to the discovery and history of radon and its potential health effects.

Marx, Jean. "Zeroing in on Individual Cancer Risk." *Science* 253 (August 9, 1991): 612-616.
Classical epidemiological methods can help identify groups of people that have a high risk of developing certain cancers, but an individual's risk varies widely depending on genetic and other factors. Researchers are trying to identify the "biomarkers" that may help explain why one individual is more susceptible than another.

Gray, George M., and John D. Graham. "Regulating Pesticides." Ch. 9 in *Risk vs. Risk: Tradeoffs in Protecting Health and the Environment*. Edited by John D. Graham and Jonathan B. Wiener. Cambridge, MA: Harvard University Press, 1995, pp. 173-192.
The authors argue that EPA's current strategy of regulating one pesticide at a time is ineffective because it fails to encourage policymakers to consider whether the substitute pesticide may be more harmful than the regulated one. This policy, in effect, trades one risk for another. For example, EPA focuses disproportionately, they argue, on whether a pesticide may be carcinogenic and less on other health effects, such as reproductive toxicity.

Lichter, Robert, and Stanley Rothman. *Environmental Cancer: A Political Disease*. New Haven: Yale University Press, 1999.
The authors use survey data to illustrate the difference between views held by cancer researchers and by activists. The authors criticize reporters for citing activists rather than the scientific community when covering stories about environmental cancer.

ECOTOXICOLOGY

Ecotoxicology is the study of the harmful effects of chemicals upon ecosystems. In natural environments, living things (plants, animals, and microorganisms) interact in many complicated ways with non-living things (air, water, and soil). Environmental toxicologists trace the movement of chemicals through air, soil, and water and assess the effect of substances, not just on human health, but on individuals, populations, and communities within an ecosystem. Some substances, such as DDT, which do not pose a significant threat to humans, may harm some animals. Other substances are a concern because of their persistence in the environment.

Francis, B. Magnus. *Toxic Substances in the Environment*. New York: John Wiley & Sons, Inc., 1994, pp. 133-136.
Francis describes the ecological and human health effects of pollutants and how they are assessed.

Society of Environmental Toxicology and Chemistry (SETAC). "Ecological Risk Assessment Technical Issue Paper, 1997."
This publication provides a basic overview of how an ecological risk assessment is conducted. Available at *http://www.setac.org/tips.html*

The University of Edinburgh: Environmental Toxicology and Ecotoxicology
These pages, located on the University of Edinburgh's *Introduction to Toxicology* site, give an overview of ecotoxicity, illustrating the discussion with descriptions of how chemical interactions, climatic conditions, and bioaccumulation affect the toxicity of substances in the environment. *http://www.med.ed.ac.uk/hew/tox/enviro.html*

U.S. Environmental Protection Agency. *Guidelines for Ecological Risk Assessment*. No. EPA/630/R-95/002F, 1998.
This publication from the National Center for Environmental Assessment provides guidelines on conducting ecological risk assessment for the Environmental Protection Agency. Available at *http://www.epa.gov/ncea/ecorsk.htm*

EPA Persistant, Bioaccumulative, and Toxic (PBT) Chemicals Initiative
This site houses information, from the Environmental Protection Agency, on PBT (persistant, bioaccumulative, and toxic) chemicals, including a special section on priority PBTs such as DDT, hexachlorobenzene, PCBs, and dioxin. There are overviews describing what each chemical is, where it can be found, how we are exposed to it, and the potential negative effects it poses. *http://www.epa.gov/opptintr/pbt/index.html*

National Research Council. *Hormonally Active Agents in the Environment*. Washington, DC: National Academy Press, 1999.
Controversy has arisen concerning substances in the environmental that may affect the endocrine system of organisms. This report, from the National Research Council's Com-

mittee on Hormonally Active Agents in the Environment, explains what is currently known about endocrine disrupters. To order see *http://www.nap.edu*

Legislation and Policy

The Toxic Substances Control Act (15 USC sec. 2601)
This act tracks and regulates the industrial chemicals produced or imported into the United States. *http://www4.law.cornell.edu/uscode/15/ch53.html*

Toxics Release Inventory
The Environmental Protection Agency requires annual reports of toxic chemical releases to the environment under Section 313 of the Emergency Planning and Community Right-to-Know Act. Information about the program and how to access the data is available through EPA's Office of Pollution Prevention and Toxics.
http://www.epa.gov/opptintr/tri/

National Council for Science and the Environment: Regulatory Reform and Risk Assessment
This site lists a number of Congressional Research Reports regarding risk assessment, including such topics as the costs of environmental protection, the use of benefit-cost analysis, and environmental health and safety tradeoffs.
http://www.cnie.org/nle/crsrsk.html

Zeckhauser, Richard, and Kip Viscusi. "Risk Within Reason." ***Science*** **248 (May 4, 1990): 559-563.**
The authors argue that an imbalance in regulatory policy results when society overreacts to a potential risk, such as a newly discovered carcinogen, that gains the attention of a political community, while other preventable risks, such as lifestyle changes, are neglected.

Freeman, Myrick, and Paul Portney. "Economics Clarifies Choices about Managing Risk." Reprinted in ***The RFF Reader in Environmental and Resource Management.*** **Edited by Wallace E. Oates. Washington, DC: Resources for the Future, 1999.**
The authors introduce readers to basic concepts of risk, economics, and regulatory policy. Freeman and Portney contend that although benefit-cost analyses can be useful, we sometimes have no more than a well-educated guess as to how physical, biological, and engineering systems operate and the economic factors that set their monetary value. Ordering information is available at *http://www.rff.org*

Labs and Activities

Chemical Education for Public Understanding Program. ***Risk Comparison.*** **New York: Addison Wesley, 1990.**
Using student activities, this teacher's manual gives an introduction to the concept of probability and the science of epidemiology. Case studies help students analyze daily personal risk decisions. The material, however, emphasizes perception of relative risks rather than evidence-based rankings.

***Understanding Environmental Health Risks: Teacher's Manual*. Science Education for Public Understanding Program, University of California, Berkeley.**
This teacher's manual for a course on environmental health risks includes ten activities on topics such as the use of chlorine to disinfect drinking water, scientists' sampling techniques, basic concepts of toxicology, and quantitative comparisons of exposure. To order call 1-516-737-1133.

National Science Teachers Association. *Teach with Databases: Toxics Release Inventory.* Arlington, VA: National Science Teachers Association, 1998.
This program, jointly developed by the National Science Teachers Association and the U.S. Environmental Protection Agency (EPA), uses the EPA's Toxics Release Inventory (TRI) as the focus for a series of activities. The teacher's guide uses this database (which is included on a CD-ROM) for activities such as using the Law of Conservation of Mass to determine the amount of chemicals released in the environment and performing a risk analysis based on local toxic releases. For ordering information call 1-800-722-NSTA.

Ost, David. "Perceived Risk." *The American Biology Teacher*, April 1995, pp. 235-238.
The authors describe an exercise in which students are asked to rank hazards individually and then as members of a group. The students are then asked to assign a numerical value to each item based upon the degree of certainty of each risk.

Ost, David. "Risk Assessment: Implications for Biological Education." *The American Biology Teacher*, March 1995, pp. 140-144.
The author suggests that students could perform sophisticated risk assessments, but notes that they must first be introduced to quantitative analysis techniques used in this field.

***EcoChallenge: An EcoRisk Simulation Game*. American Industrial Health Council, 1999.**
EcoChallenge is an interactive board game that demonstrates the principles and challenges of conducting an ecological risk assessment. Students investigate scenarios including: contaminated soil and groundwater at a petroleum marketing terminal; approval of a chemical for use in personal care-products; and evaluating the risks of a new industrial chemical. This may be a useful exercise to help students understand the complexities of environmental decisionmaking. See *http://www.aihc.org* for more information.

Additional Resources

Tomes, Nancy. *The Gospel of Germs*. Cambridge, MA: Harvard University Press, 1998.
Only one hundred years ago, most people would have scoffed at the idea that diseases were caused by tiny microorganisms. They shared toothbrushes, drank unfiltered water, and stored meat without any thought of consequences to health. This book provides a fascinating history of the change in public health brought by the recognition of the role germs play in transmitting disease.

Monmonier, Mark. *Cartographies of Danger: Mapping Hazards in America*. Chicago: University of Chicago Press, 1997.
The author shows how geographic mapping techniques can be used to model risks from natural sources, such as earthquakes and floods, and from human sources, such as electromagnetic fields and radon.

Harr, Jonathan. ***A Civil Action.*** **New York: Random House, 1995.**
This best-selling book recounts the costly and lengthy lawsuit filed against several corporations on behalf of a group of Woburn, Massachusetts, residents stricken with leukemia. The suit, eventually settled out of court, alleged that the corporations were responsible for contaminating the local water supply. Ohio State University Geological Sciences Department has an excellent site, "Science in the Courtroom," which includes the geological and hydrological evidence upon which the parties to the suit relied.
http://www.geology.ohio-state.edu/courtroom/

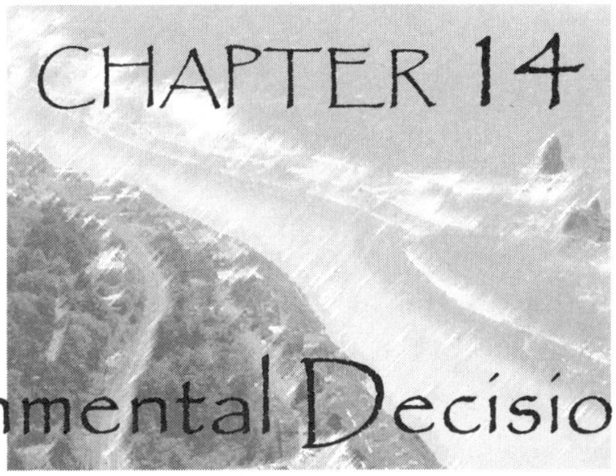

CHAPTER 14

Environmental Decisionmaking

> The difficulty of converting scientific findings into political action is a function of the uncertainty of the science and the pain generated by the action.
>
> ★William Ruckelshaus in **Managing Planet Earth**★

Making decisions about how best to conserve natural resources and environmental quality is not a simple task, particularly since many environmental problems are not easy to mitigate and the solutions may be expensive and pose other risks. Science informs decisions by providing information about natural processes, but there are often competing interests, values, and risks that must be considered. In any case, policymakers are often pressed to make decisions about questions in which there are areas of scientific uncertainty. Inevitably there are tradeoffs that must be considered. Concerns about municipal water quality, for example, could be addressed if local governments purchased laboratory-tested bottled water for every household. The cost, though, would be enormous and would reduce funding available for other public services. In some areas, wildlife managers would like to remove introduced species that they regard as a nuisance and that compete with other species in the ecosystem; others argue that these new species have become an integral part of a new ecosystem and should remain. For almost every environmental issue, there are tradeoffs that must be weighed.

To assist policymakers in weighing the various considerations that are inevitable in making decisions about environmental issues, a quantitative analysis of the factors is often helpful. For example, risk assessment is a means of distilling scientific information so that it is useful for making decisions. Economic analysis is another tool that helps policy makers assess the relative benefits and costs that would accrue if one strategy is chosen over another. There are many other considerations that must be weighed in making public decisions, such as the impact of those decisions on various communities and ecosystems. The following resources discuss some of the factors that influence environmental decisionmaking.

ECONOMICS AND THE ENVIRONMENT

No matter how rich we are, we do not have the resources to achieve all worthwhile goals, so we must make choices. Economics is the subject that deals with the allocation of scarce resources among competing aims. It is a tool of analysis that enables policymakers to assess the consequences of choosing one action over another. Decisionmaking depends on four key economic concepts:

1. **Opportunity costs**
To the extent that society devotes resources to one area, it foregoes the opportunity to devote them to another. Indoor pollution, for example, currently poses a far greater risk than outdoor pollution, partly because most of us spend more than 90 percent of our time indoors. Smoking and other indoor pollutants often result in levels of pollution much higher than they are outdoors. Yet the amount we spend on indoor quality is so minuscule that no one has ever bothered to calculate it. By contrast, we spend $30 billion a year on outdoor air quality.

2. **Benefit-cost analysis**
If the benefits of taking an action are less than the costs, the action may not be warranted. For example, the EPA may ban a new pesticide because it is thought to cause harm to certain animals. Farmers might continue to use an older pesticide, which may cause more damage than the new substance would have inflicted. A benefit-cost analysis provides information that assists policymakers in weighing the consequences of choosing between alternative policies.

3. **Tragedy of the Commons**
In absence of property rights, there are few incentives to conserve resources over the long term (Hardin, 1968). Historically, in England and New England towns, the commons was the area where farmers grazed their cattle. Each farmer brought more cattle to graze until eventually the commons were overgrazed. When land or other resources are owned in common, each has an incentive to take as much from the common share as possible, because otherwise, others will take more. The oceans are an example of a commons, and the problems that may result when resources are shared. Each fisherman has an incentive to catch as many fish as possible, since others will do the same, which can lead to overfishing and depletion of the fish population.

What is GDP?

A term that is often used, and rarely explained, is GPD or gross domestic product. The gross domestic product is the sum of all the goods and services produced by a country. It is used as an indicator of the condition of a country's economy and is useful as a means of comparing the country's economic performance over time, and to the economic performance of other countries. The GDP of the United States is estimated quarterly by the Department of Commerce, Bureau of Economic Analysis. The Bureau's web site includes more information about GDP and how it is calculated **(http://www.bea.doc.gov/)**. The National Council on Economic Education has an online lesson, "Gross Domestic Pizza," that explores how GDP is measured **(http://www.economicsintl.org/pubs/lessons/nov99/nov99.html)**.

4. **The importance of incentives**
Individuals (or corporations or nations) who have something to gain, financial or otherwise, by taking a certain action, will do so readily, and will avoid taking actions that are not in their interests, if at all possible. Traditionally, environmental policies have prescribed certain technologies or actions rather than providing incentives to achieve the goal. Under the 1977 amendments to the Clean Air Act, for example, the government prescribed technologies that power plants were required to use, such as state-of-the-art scrubbers. This proved very costly, so electric utility companies decided to keep older, more polluting power plants in service rather than build expensive new plants with the required technology. To counteract this problem, subsequent legislation set up a market in sulfur oxide emissions to provide incentives for reducing emissions, an approach which has proven successful (See Air Quality).

Most would agree that actions that maximize the benefits to be gained are preferable. Disagreements arise over how to quantify benefits so they can be compared with the costs. What dollar value, for example, should be used for a small endangered fish? Economists argue that, even though it may be difficult to value many intangibles, policy makers must do so at least implicitly to make choices. If we build a hospital on land that is the only known habitat of the Delhi sand fly, then we must make a choice about the value of the hospital and the value of the fly.

There is some disagreement among economists concerning the extent to which market prices reflect all the costs involved in a transaction. If oil is extracted, for example, the remaining stock of oil in the world is diminished. Some argue that this loss is not included in the price of the oil and not counted in the national income statistics. Classical economists would say that investors and speculators calculate whether the remaining stocks are becoming depleted. If the stocks are being significantly reduced, some people will retain a supply to sell at a higher price in the future. This will have the effect of driving up current prices and spurring the demand for alternate technologies. Thus, in a well-functioning market, prices do reflect future scarcity.

It is true, however, that an economic system does not reflect all costs. A factory that dumps smoke into the air imposes the costs of polluted air on nearby towns. Costs or benefits that are not reflected in the price of goods and services are referred to as *externalities*. One approach suggested for environmental policy is to include the costs of externalities in prices through measures such as taxes. Increasing taxes on gasoline, for example, makes fuel more expensive and encourages individuals to drive fewer miles. This approach has proven difficult to implement for various reasons, including public opposition to higher taxes and the problem of determining quantitative measures for certain things such as aesthetic values, or of the value of services provided by global environmental systems.

There are several economics textbooks noted below that provide a basic introduction to economics, in addition to several online tutorials that explain basic economic concepts in clear terms. Almost every discussion of environmental issues in the media refers to economic terms such as opportunity costs, externalities, diminishing marginal returns, gross domestic product, and benefit-cost analysis. The textbooks and online tutorials will be helpful in reviewing the meaning of these terms.

Field, Barry. *Environmental Economics: An Introduction*, New York: McGraw-Hill, 1994.
This is a basic college-level economics textbook. The author uses a number of environmental issues to illustrate how economic principles and analysis techniques are applied to environmental problems. This book may be available in a public or a university library.

Siebert, Horst. "The Political Economy of Environmental Scarcity." Ch. 9 in *Economics of the Environment: Theory and Policy*. Third Edition. New York: Springer-Verlag, 1992, pp. 151-157.
This is a more advanced college-level economics textbook. This chapter is helpful in understanding the relationship between market prices and supplies of resources.

Prato, Tony. "Property Rights and Externalities." Ch. 5 in *Natural Resources and Environmental Economics*. Ames, IA: Iowa State University Press, 1998, pp. 91-110.
Chapter 5 of this college-level text explains the effects of property rights and externalities on markets and resource allocation. The chapter includes a discussion of commonly-held property and property rights.

Cyber-Economics
A complete online, two-semester course in introductory economics, this is one of the most comprehensive basic economics site on the web.
http://ingrimayne.saintjoe.edu/econ/TitlePage.html

EconEd Link: K-12 Economics Resources
EconEd Link is an educational site devoted to teaching economics. Primarily for teachers, the site has lesson plans, data, and links focusing on the economics behind current issues.
http://www.economicsamerica.org/econedlink/

Environmental Protection Agency: Economy and Environment
This EPA site includes a publication on "Environmental Economics in Plain English," and links to other sites dealing with economics and the environment.
http://www.epa.gov/economics/

Hardin, Garrett. "The Tragedy of the Commons." ***Science*** **162 (December 13, 1968): 1243-1248.**
In this classic article, Hardin present the problem of shared resources. A herdsman benefits by maximizing the number of cattle he has grazing on a common pasture. But, the pasture will be of no benefit to any herdsman if it has been overgrazed. Thus, in the absence of mechanisms to protect public resources, individuals will act in self-interest without regard to collective good.

Farrow, Scott, and Michael Tolman. "Using Benefit-Cost Analysis to Improve Environmental Regulations." ***Environment*****, March 1999, pp. 12-15, 33-37.**
This article offers a very clear explanation of the role of economic analysis in policymaking. In particular, the section, "A Primer on Discounting," is an easy to comprehend discussion of an important concept.

Moore, John. "Cost-Benefit Analysis: Issues in Its Use in Regulation." Congressional Research Service Report for Congress, No. 95-760 ENR, June 28, 1995.
This report, written for members of Congress, provides a clear explanation of what benefit-cost analysis is and how it is currently used in regulation decisions, using the Environmental Protection Agency's analysis of proposed air quality regulations as an illustrative example. Available at *http://www.cnie.org/nle/rsk-4.html*

Blodgett, John E. "Environmental, Health, and Safety Tradeoffs: A Discussion of Policymaking Opportunities and Constraints." Congressional Research Service Report for Congress, No. RL30043, 1999.
Tradeoffs are inevitable in policymaking. This briefing for Congress and staff discusses how decisions by a legislator or other policymaker at a particular time and place are constrained by institutional structure and rules and describes the implications of cost-benefit analysis and risk assessment in the context of congressional and administrative decision-making. Available at *http://www.cnie.org/nle/rsk-13.html*

Meyers, Norman, and Kent, Jennifer. ***Perverse Subsidies: Tax $s Undercutting our Economies and Environments Alike*****. Manitoba, Canada: International Institute for Sustainable Development, 1998.**
This extensive study attempts to calculate the economic and environmental costs of "perverse subsidies," that is, government policies that subsidize activities that have detrimental effects on natural resources. As one example, the authors point to the U.S. where

"one government agency heavily subsidizes irrigation for crops that another agency has paid farmers not to grow." While the estimates may be high (or low), the report is useful in explaining how government policies can have unintended consequences. This report (which is very lengthy) can be downloaded from *http://iisd1.iisd.ca/*

Sustainable Development

There is an ongoing debate about the relationship between economic growth and the environment. Some believe that because economic growth means more people will consume more goods, they will use more energy and generate more waste and pollution. Others argue that "richer is cleaner," and that it is only when a society has moved beyond subsistence level and reached a certain level of wealth that it can afford the technologies to reduce emissions and clean up wastes. Moreover, people in developing countries are desperately trying to improve the quality of their lives. Sustainable development is a term that has been used in the past two decades to recognize the need to reduce poverty through economic growth while preserving ecosystem health and the long-term stock of natural resources. There are disagreements about what sustainable development means and how to strike the balance between human needs and environmental protection. These resources will help the reader gain an understanding of issues related to economic growth and sustainable development.

MacNeill, Jim. "Strategies for Sustainable Economic Development." ***Scientific American*****, September 1989, pp. 155-165.**
MacNeill summarizes the findings of the World Commission on Environment and Development (or the Bruntland Commission, as it is more commonly known) on how to balance economic growth and the environment. McNeill, who served as secretary general of the Commission, reviews the perverse subsidies, political, and social factors that have hindered development in poor countries.

Ruckelshaus, William. "Toward a Sustainable World." ***Scientific American*****, September 1989, pp. 166-174.**
Ruckelshaus, who was the first director of the Environmental Protection Agency, argues that environmental protection and economic development are complementary, not antagonistic policies.

Rosenberg, Nathan, and L. Birdzell Jr. "Science, Technology and the Western Miracle." ***Scientific American*****, November 1990, pp. 42-54.**
The authors address why long-term economic growth has occurred in the West while developing countries lag behind in per capita income. They conclude that the success of science and its application to technology has fueled the "Western miracle," and that a market economy and independent sources of decisionmaking provide an impetus for innovation.

Daly, H. ***Steady-State Economics*****. Washington, DC: Island Press, 1991.**
Daly rejects the idea that economic growth is consistent with protecting the environment and argues that taxes should be shifted from income to consumption of natural resources.

Political Economy Research Center
This non-profit organization provides online publications that discuss market-based approaches to environmental issues. *http://www.perc.org/*

Repetto, Robert. "Accounting for Environmental Assets." ***Scientific American*****, June 1992, pp. 94-100.**
Repetto argues that the United Nations System of National Accounts (SNA) is outdated because consumption of natural resources is not included, even though a loss of natural capital may lead to a decrease in future production. Because the economies of many developing countries are dependent upon natural resources, the account data, he maintains, conceals future scarcity.

Arrow, Kenneth et al. "Economic Growth, Carrying Capacity, and the Environment." ***Science*** **268 (April 28, 1995): 520-521.**
The authors note that the relationship between economic growth and environmental quality is not a simple one. Wealthy countries can afford to spend money cleaning up local environmental problems, but reductions in pollutants in one country can be offset by an increase in pollutants in another, if manufacturing facilities are relocated to countries with lower environmental standards. They argue that economic growth is not a substitute for institutional reforms that tend to protect natural resources.

Environmental Justice

In 1982, a proposed landfill in Warren County, North Carolina was protested by residents of an African-American community that was located close to the planned facility. Some observers asserted that local officials had been practicing a form of environmental racism, citing studies that hazardous waste treatment, storage, and disposal facilities are disproportionately located in areas occupied predominantly by minorities. Others argued that poor communities are often ill-represented in political circles and therefore are the easiest places to locate objectionable facilities. Eventually the EPA granted the permits for this waste disposal site, which brought nationwide protests.

In response to protests stemming from this and similar incidents, President Clinton, on February 11, 1994, issued an executive order (EO 12,898) mandating that the Environmental Protection Agency establish an Office of Environmental Justice. In April 1998, the EPA defined environmental justice as fair treatment, meaning that "no group of people, including racial, ethnic, or socioeconomic groups, should bear a disproportionate share of the negative environmental consequences resulting from industrial, municipal, and commercial operations or the execution of federal, state, local and tribal programs and policies." According to the EPA definition, deliberate discrimination need not be involved. Any siting which disadvantages protected groups more than proportionately is in violation of EPA's rules. The difficulty arises in defining what to measure and in comparison to what.

As a first step in evaluating whether a group is unduly disadvantaged, a policy maker must consider who is affected. Most ethnic data relate to census tracts, zip codes, city boundaries, and counties. If the facility is to be located in a wealthy county, but located near the county border close to a poor community, how does the policy maker draw the line? Should prevailing winds be considered? Many industrial sites are located where land is cheap; people of low income may choose to live in those areas to minimize living expenses. How are these decisions weighed?

Another difficulty arises in determining whether and how particular groups will be disadvantaged. Landfills, chemical plants, and other industrial works bring benefits to some although they may harm others. They create jobs, change land values, and generate revenues that are spent in the community. How do officials compare the benefits with the losses? How should potential health risks from a facility be compared with the overall health benefits that jobs and higher incomes bring?

There are conflicting studies of the question of disparate impacts. A study in Massachusetts and a recent GAO report, for example, found no disproportionate impact by ethnicity; a study in Chicago found that industrial sites were located close to high-income areas. Decisions about siting of industrial and waste facilities inevitably generates controversy. The following resources address some of the issues involved.

Zimmerman, Rae. "Issues of Classification in Environmental Equity: How We Manage is How We Measure." ***Fordham Urban Law Journal*** **21 (November 3, 1994): 633-669.**
This article provides a detailed overview of the studies that have been done in this area. Results of studies have varied according to the size of the geographic area included in the comparisons. This article presents an excellent analysis of the issues, but it may be difficult to find except in a law library.

Steinberg, Michael. ***Making Sense of Environmental Justice, Briefly: Perspectives on Legislation, Regulation, and Litigation.*** **National Legal Center for the Public Interest, June 1999.**
In this monograph, Steinberg critically examines some of the early studies showing higher rates of exposure to hazardous waste sites by ethnicity. He discusses the division within some poor communities between those opposing industry because of perceived pollution and those who hope to attract industry for economic benefit. The full text is available at *http://www.nlcpi.org/books/briefly35/toc.htm*

U. S. Environmental Protection Agency: Environmental Justice
This site includes access to the *EJ Quarterly* and National Environmental Justice Advisory Council documents. The site also provides links to related information such as brownfields, or abandoned industrial sites, often located in inner city areas.
http://www.epa.gov/swerosps/ej/

The Environmental Justice Information Page
Some groups have defined Environmental Justice more broadly as encompassing anything that might adversely affect people of color, the poor, native Americans, all species, the environment and the political, economic, cultural and environmental self-determination of all peoples. This site developed by University of Michigan students has an extensive list of papers and other resources on environmental justice, including a list of "Principles of Environmental Justice" developed at the People of Color Environmental Summit in 1991. *http://www-personal.umich.edu/~jrajzer/nre/index.html*

Bullard, Robert et al. "Sprawl Atlanta: Social Equity Dimensions of Uneven Growth and Development." Environmental Justice Resource Center, January 1999.
This study includes an analysis of variables that have contributed to social inequities resulting from urban sprawl and unplanned growth in Atlanta. An executive summary of this report is available at *http://www.ejrc.cau.edu/*

Foreman Jr., Christopher. "...And Environmental Justice for All?" *Priorities*, 1997.
This Brookings Institution scholar reviews the conflicting evidence about the relationship between race and the likelihood of living near a commercial hazardous-waste site. Available at *http://www.brook.edu/views/articles/foreman/1997priorities.htm*

Environmental Impact Analysis

Another consideration that affects decisions by local, state, and federal officials is the potential impact of a policy on the environment. One provision of the National Environmental Policy Act requires agencies of the federal government to conduct analyses on all of their proposed projects and programs "which may have a significant impact on the environment," and to make the results of these analyses available to planners and decisionmakers in both the public and private sectors, and to the public in general, in the form of Environmental Impact Statements. Now, state and local governments generally require that environmental impact analyses be conducted for most major development projects.

> If the individual has a warm personal understanding of land, he will perceive of his own accord that it something more than a breadbasket. He will see land as a community of which he is only a member, albeit now the dominant one. He will see the beauty, as well as the utility, of the whole, and know the two cannot be separated. We love (and make intelligent use of) what we have learned to understand.
>
> ★Aldo Leopold, **The River of the Mother of God and Other Essays by Aldo Leopold★**

The National Environmental Policy Act requires that agencies of the federal government "insure that presently unquantified environmental amenities and values are given appropriate consideration in decisionmaking, along with economic and technical considerations." The task is formidable because it often calls for making projections with incomplete information and often reveals inadequate understandings of interrelationships among many variables. Methods of assessing environmental impact are complex and are often controversial because they call for mixtures of objective and subjective analysis, and some values and considerations are difficult to quantify.

An environmental impact statement generally includes factors such as the geology, soils, and hydrology of the site; the project's effect on air quality, wild life, vegetation, and water quality; plus its effect on noise levels, public services, employment, housing, traffic, transit, land use, cultural resources, aesthetics, and quality of life. These analyses demonstrate the interrelationships among scientific, social, and economic considerations, and the complexities and uncertainties involved in making choices. The following resources provide more detail about environmental impact analysis.

Botkin, Daniel B., and Edward A. Keller. *Environmental Science: Earth as a Living Planet*. Second Edition. New York: John Wiley & Sons, Inc., 1998.
Chapter 29, "Environmental Impact and Planning," of this college-level environmental science text provides a clear overview of the environmental impact analysis process.

Jain, R. K., et al. *Environmental Assessment*. New York: MCGraw-Hill, 1993.
This book is intended to guide engineers through the process of an environmental impact analysis. The procedures for assessing economic and social impact of projects are described in detail. This is an expensive volume, but it may be available in public or university libraries.

Council for Environmental Quality: CEQ NEPANet
The Council's site includes guidelines for preparing an environmental impact statement, the full text of the National Environmental Protection Act and regulations, and data on the number of Environmental Impact Statements filed each year. Copies of statements filed by federal agencies can also found here. *http://ceq.eh.doe.gov/nepa/nepanet.htm*

Harrop, D. Owen, and J. Ashley Nixon. *Environmental Assessment in Practice*. New York: Routledge, 1999.
This is a scholarly introduction to environmental assessment.

University of California--Santa Cruz, Environmental Studies 151: Links to Environmental Documents
This university environmental science course site has compiled links to environmental assessment documents on the web, including assessments of controversial projects by the National Park Service, the Bureau of Land Management, and the Department of the Interior, such as the statement on the impacts of the reintroduction of the Mexican wolf in the Southwestern United States. *http://wwwcatsic.ucsc.edu/~envs151/lienv.htm*

National Environmental Policy Act of 1969 (Public Law 91-516) as amended by Public Law 94-52 (1975), Public Law 94-83 (1975), and Public Law 99-160 (1985).
The text of the original Act and its revisions, setting forth both a rationale for its passage and a list of specific requirements for agency personnel. Available on the CEQ NEPANet site above.

Westman, Walter E. *Ecology, Impact Assessment, and Environmental Planning*. New York: John Wiley & Sons, Inc., 1985.
The text, written by a biologist, explains methods of predicting human impacts on ecosystems.

Lemons, John, ed. *Scientific Understanding and Environmental Problem Solving*. New York: Blackwell Science, 1995.
This collection of articles assesses current scientific understanding of issues such as water quality, ecosystem health, nuclear waste, and fisheries management and addresses the implications of environmental decisionmaking in the face of scientific uncertainties. This is an expensive volume; it may be available as a reference book in public or university libraries.

LAND USE

Public decisions that affect how land may be used are often controversial. Land use is a term used to describe the various ways in which human beings make use of the land and its resources, including farming, mining, building, or grazing sheep and cattle. Choices about how land is used are made by those who own or control the land, but these choices are limited by the physical and biological characteristics of the land, including its climate, soils, and geography. Lands with rich soils are most suitable for farm-

"Land is the only thing in the world that amounts to anything...for 'tis the only thing in the world that lasts, and don't you be forgetting it! 'Tis the only thing worth working for, worth fighting for–worth dying for."

★Margaret Mitchell, **Gone with the Wind**★

ing; lands that are easily flooded are less suitable for housing than land at higher elevations. Economic considerations also affect land choice decisions. For example, if it is too costly to build on a plot of land because of its elevation or incline, another use may be chosen.

Land use choices are also limited by institutional factors and, in the United States, land use is governed by a mix of local, county, state, and federal laws. Although most of the land in the United States is privately owned, the federal government is the single largest owner of land, administering about 32 percent of the land in the U.S., mostly in the Western states. There have been major changes in patterns of land use in the United States since its founding. In 1790, for example, over 90 percent of the U.S. population lived on farms; now less than 3 percent of the population do.[33] Although many acres of forest were cleared in the 18th and 19th centuries for cropland, that trend is reversing and regeneration of vegetation is occurring in many areas.

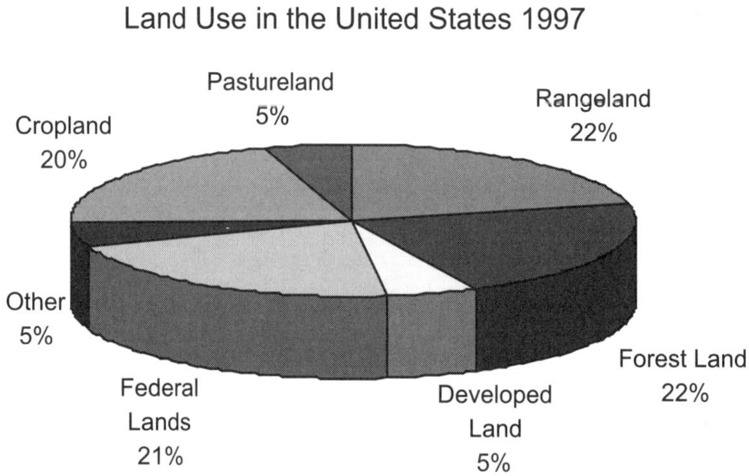

Source: U.S. Department of Agriculture, Natural Resource Conservation Service, 1997, **Natural Resource Inventory.**

Analysis of land use and land cover has been greatly facilitated in the last few decades by the availability of new technologies, including remote sensing. The development of geographic information systems (GIS) has made it possible to combine physical characteristics of the land with information about population, economic data, and other information for further analysis. The following resources may be helpful in researching land use in the United States.

Meyer, William B. "Past and Present Land Use and Land Cover in the USA." ***Consequences*, Spring 1995, pp. 24-33.**
This article provides an excellent summary of how land is currently used in the United States and the changes in land use that have occurred since colonial times. He notes that the term "land reclamation" was originally used to refer to converting land for use by clearing vegetation or irrigating it (as the U.S. Bureau of Reclamation was created to do); now it is more likely to be used to describe restoring land damaged by human use.
http://www.gcrio.org/CONSEQUENCES/spring95/

U.S. Department of Agriculture: 1997 Natural Resources Inventory
This survey, conducted every five years, details data from 800,000 statistically-selected locations on land cover, land use, soil erosion, prime farmland soils, wetlands, habitat

diversity, selected conservation practices, and other natural resource information. *http://www.nhq.nrcs.usda.gov/NRI/1997/*

U.S. Geological Survey Biological Resources Division. ***The Status and Trends of the National's Biological Resources.***
This report has an extensive chapter on land use in the United States. Available at *http://biology.usgs.gov/s+t/SNT/*

U.S. Geological Survey: The National Atlas of the United States of America
This online atlas combines geological data, including soils, terrain contours, volcanoes, and principal aquifers, with statistical data on population distribution, land cover, and other information. This information can be used to examine characteristics of geographic areas using different methods, such as county, state, and zip code boundaries, or geographic coordinates like latitude and longitude. *http://www.nationalatlas.gov/*

National Wilderness Institute: State by State Government Land Ownership
This site lists the percentage of land in each state owned by the state and the federal governments. *http://www.nwi.org/Maps/LandChart.html*

Leopold, Aldo. ***A Sand County Almanac.*** **New York: Ballatine Books, 1982.**
In this series of nature essays, Leopold articulates an elegant statement of the appropriate relationship between humans and the land.

Environmental Systems Research Institute, Inc. ***Getting to Know ArcView GIS.*** **Cambridge, UK: GeoInformation International, 1997.**
This manual provides step-by-step instructions for using the ArcView geographic information system (GIS), which permits users to combine geographic and other data, such as vegetation patterns, rainfall amounts, land use data, for analysis. The attached CD-ROM includes a hands-on tutorial as well as a demo copy of the software. Data sets can be obtained from a number of sources including the U.S. Geological Survey (*http://www.usgs.gov/research/gis/title.html*) and the Environmental Systems Research Institute (*http://www.esri.com*).

Environmental Systems Research Institute: GIS for Everyone
This site introduces geographic information systems (GIS), permits user to download free GIS software tools, and download GIS information.
http://www.esri.com/gisforeveryone/index.html

Association of American Geographers: Human Driving Forces and their Impacts on Land Use/Land Cover
This is an extensive online curricula on human impacts on land use and land cover created as part of the Virtual Geography Department, an Internet project to link geography departments around the world. *http://www.aag.org/HDGC/www/LUCC/titlepage.html*
Additional curricula resources are available on the Virtual Geography Department's site at *http://www.utexas.edu/depts/grg/virtdept/contents.html*

Woodrow Wilson Environmental Science Institute Rutgers: Observing Earth from Space
This online guide is designed to provide teachers and students with access to learning modules, information, resources and references on the topics of remote sensing, mapping, and GIS. *http://www.woodrow.org/teachers/esi/1998/r/remote/*

Universities Space Research Association: Earth System Science Online
This site is designed as a repository for Earth System Science education and research resources. Links are included to digital mapping data, software, and educational tools. *http://www.usra.edu/esse/essonline/*

Environmental Protection Agency: EnviroMapper
This site allows the user to map several types of environmental information, including drinking water, toxic and air releases, hazardous waste, water discharge permits, and Superfund sites by geographic area. *http://www.epa.gov/enviro/html/em/index.html*

Monmonier, Mark. *Cartographies of Danger: Mapping Hazards in America.* Chicago: University of Chicago Press, 1997.
The author shows how geographic mapping techniques can be used to model risks from natural sources, such as earthquakes and floods, and from human sources. He illustrates how land use planners in California could reduce damages suffered during earthquakes by mapping technologies.

Wessels, Tom. *Reading the Forested Landscape: A Natural History of New England.* Woodstock, VT: The Countryman Press, 1997.
The authors uses the forests in his native New England to demonstrate how to find evidence for human activities and disturbances such as fire, logging, pastures, and blight, that can be used to trace the history of how land has been used.

Legislation and Policy

Policy.com: Public Lands
This site includes papers and articles from a variety of sources relating to use of public lands, urban sprawl, smart growth, and other land use issues. *http://www.policy.com/issues/14/item70.asp*

Conniff, Richard. "Federal Lands: New Showdowns in the Old West." *National Geographic*, February 1994, pp. 2-39.
Conflicts over land use between loggers, ranchers, and environmental groups have been heated in the West. The author visits Salmon, Idaho, a small Western town, to "go behind the bitter name-calling," of western land disputes. This is an even-handed look at the issues that divide Westerners.

Meltz, Robert. "The Property Rights Issue." Congressional Research Service, No. 95-200 A, January 20, 1995.
One of the most contentious issues in environmental policy in the last two decades is the debate over the protection of property rights and the "Takings Clause" of the Constitution. This briefing paper for Congress provides a good overview of the issues involved, a brief history of the relationship between public goals and private rights, and the pros and cons of property rights legislation. *http://www.cnie.org/nle/econ-11.html*
Links to site with additional information on this issue from a variety of sources can be found at *http://dmoz.org/Society/Issues/Property_Rights/Regulatory_Takings/*
A list of recent cases relating to the takings issue can be found at *http://www.cnie.org/nle/econ-12.html*

History and the Land

Runnels, Curtis. "Environmental Degradation in Ancient Greece." ***Scientific American*****, March 1995, pp. 72-75.**
Archaeologists are using new tools, such as remoting sensing, to detect the effects of past human activity. This article reviews the findings of archaeological research in Greece, in which investigators have uncovered evidence of land abuse and erosion in ancient times.

Mann, Charles. "Earthmovers of the Amazon." ***Science*** **287 (February 4, 2000): 786-789.**
This article discusses the debate among archaeologists over whether a major prehistoric civilization existed in the Bolivian Amazon area. Researchers have discovered evidence of extensive earthworks that seem to be the remains of transportation canals, causeways, and croplands. Others argue that the nutrient-poor soils of the rain forests in Amazon made the land unsuitable for support of a major civilization.

Russell, Emily W. ***People and the Land through Time: Linking Ecology and History.*** **New Haven: Yale University Press, 1998.**
All human actions have some effect on the landscape, even if subtle. The author, a Rutgers University environmental geologist, explores how human history can be revealed by studying evidence of changes in ecosystems and landscapes and how changes in the environment have affected human history.

Cities: The Past, Present, and Future of Human Ecosystems

The most intensive use of land by humans is in cities. There has been a major urban shift of the population over the previous century. At the beginning of the 20th century, most of the world population lived in rural areas; now more than half live in urban areas. It is projected that within twenty-five years, more than two-thirds of the global population will live in cities (World Resources Institute, 1). There has been increased attention, therefore, to the need to study the flow of energy and materials in human ecosystems.

> A town is saved, not more by the righteous men in it, than by the woods and swamps that surround it. A township where one primitive forest waves above, while another primitive forest rots below—such a town is fitted to raise not only corn and potatoes, but poets and philosophers for the coming ages.
>
> ★Henry David Thoreau, **Walking**★

Cities are where human civilization began; the word "civilization" comes from the Latin *civitas*, meaning city. The world's great universities, libraries, cathedrals, and museums are found in cities. They have been the center of scientific discovery and technological innovation, and of commerce and literacy. The immigration of populations around the world to urban areas has been fueled in large part by hope for a better quality of life than that afforded by a labor-intensive existence in rural areas.

Cities, however, can be also be congested; even in ancient times a few large cities, including Rome and Carthage, had almost one million inhabitants (Goudie, 25). There are now 23 "megacities," which have populations of over 10 million. Many of these cities are in developing countries, which have not yet built an infrastructure for meeting the basic needs of rapidly growing populations. By one estimate, over 200 million persons in urban areas lack access to clean drinking water and more than 400 million have no access to sanitation (World Resources Institute, 1). Air quality in many large cities around the world is a serious concern.

There are environmental benefits to cities. In the United States, for example, where most of the population lives in urban areas, only five percent of the land is covered by cities, paved roads, and the built environment, leaving land not used for crops or pasture for natural habitats (Ausbel 1997, 5). Urban populations tend to have fewer children than those in rural areas since children are not required for labor (World Resources Institute, 2). It is also possible to use energy, communications, and transportation systems more efficiently in urban areas, and designs for future cities focus on increasing the effectiveness of these systems. The following resources discuss aspects of past, present, and future cities.

Botkin, Daniel B., and Edward A. Keller. *Environmental Science: Earth as a Living Planet*. New York: John Wiley & Sons, Inc., 1998.
This text includes an interesting discussion of the city as an ecosystem. The authors include several interesting case studies, including one on Venice and another entitled, "Cities and the Fall Line," which explains why so many cities were built near fall lines, the place where the elevation of the land drops abruptly and waterfalls are common.

Goudie, Andrew. *The Human Impact on the Natural Environment*. Cambridge, MA: MIT Press, 1996.
Goudie includes several chapters that assess scientific evidence of the effect of urbanization on the natural environment, including air quality, water quality, and the weather. He also notes that some wildlife, particularly deer, birds, raccoons, opossums, and insects, thrive around urban areas.

World Resources Institute. *World Resources 1998-99: A Guide to the Global Environment*. New York: Oxford University Press, 1998.
A joint effort of WRI, the World Bank, and the United Nations, this guide focuses on urban environments. The guide includes brief overviews of global trends in population, food production, deforestation, and other environmental issues, with extensive data. Information on ordering is available at *http://www.wri.org* or at 1-800-822-0504.

"Air Pollution in the World's Megacities." *Environment*, March 1994, pp. 4-13, 25-37.
This article summarizes a study of air quality in twenty of the world's largest cities by the Global Environmental Monitoring System, an environmental assessment program of the United Nations.

National Academy of Engineering. *The Bridge*, Winter 1999.
This special issue of *The Bridge* includes articles that focus on the engineering challenges of building cities. Articles include, "Wild Ideas in Future Cities," "Industrial Ecology and the Ecocity," "Sustainable Cities," and "Megacities and the Developing World." The publication is available online at *http://www.nae.edu/TheBridge*

Jacobs, Jane. *Cities and the Wealth of Nations: Principles of Economic Life*. New York: Vintage Books, 1985.
This classic book makes the argument that cities are the engines of economic growth. Jacobs notes that, like a natural ecosystem, the more diversity there is in a city, the more flexible, adaptable, and stable its economy will be.

U.S. Long Term Ecological Research Network (LTER)

This program was established in 1980 by the National Science Foundation to support research on long-term ecological phenomena in the United States. There are now 21 sites throughout the U.S., representing diverse ecosystems. Two sites, one in Phoenix and one in Baltimore, have recently been selected as urban ecology sites, designed to study the long-term relationship between ecological and social and economic factors in urban environments. These studies have only been in operation for a short period; however, each site has initial information about the urban ecosystem. *http://www.lternet.edu*

The Phoenix project site is at *http://caplter.asu.edu/*

The Baltimore site is at *http://baltimore.umbc.edu/lter/*

Herman, Robert, and Jesse Ausbel. *Cities and Their Vital Systems: Infrastructure Past, Present, and Future*. Washington, DC: National Academy Press, 1988.

This book includes articles that discuss how cities and their supporting systems, such as transportation, communications, and utilities, developed and how those systems can be improved to meet current needs. The chapter by W. Brian Arthur addresses the events, natural characteristics, or other factors that determine where urban centers are developed. This is an expensive volume; it may be possible to find it in a public or university library. It can be also be read or ordered at *http://www.nap.edu*

Ellis, Cliff. "History of Cities and City Planning." *SimCity Classic Users Manual*. Maxis, Inc., 1993.

Ellis, a professor of urban planning at State University of New York, Albany, wrote this concise history of cities and city planning for the user's manual of the popular simulation game, *SimCity*, in which users build and maintain a computer-simulated city. The essay is instructive in its overview of the constraints on city forms, the way cities are designed according to local climate, terrain, and natural transportation routes such as bodies of water, and its summary of the major components, such as transportation, industry, and infrastructure. This essay is available online at *http://www.angelfire.com/ar/corei/udlinks.html*. The *SimCity* simulation might also be a useful exercise for demonstrating a city's components.

Mumford, Lewis. *The City in History: Its Origins, Its Transformations, and Its Prospects*. New York: Fine Communications, 1997.

Originally published in the 1960s and very influential in shaping the thoughts of urban planners, this is a history from the earliest times, through medieval villages to modern cities of city life. Mumford was particularly critical of what he saw as the dehumanizing aspect of contemporary cities.

Cronon, William. *Nature's Metropolis: Chicago and the Great West, 1848-1893*. New York: W. W. Norton & Co., 1991.

This is an economic and environmental history of Chicago and its relationship to the surrounding countryside. Cronon examines how the city developed as the distribution center for the Midwest as the region became the farming and ranching center of America.

Davis, Margaret Leslie. *Rivers in the Desert: William Mulholland and the Inventing of Los Angeles*. New York: HarperPerennial, 1993.

This lengthy book tells the story of rise and fall of William Mulholland, who made it possible to build a major city in a desert by building what was then an engineering marvel—

an aqueduct still in use, the largest and longest in the country. This book is out of print but should be available in a public library.

Caro, Robert A. ***The Power Broker: Robert Moses and the Fall of New York.*** **New York: Random House, 1975.**
This is a Pulitzer-prize winning biography of Robert Moses, a powerful bureaucrat who fought Tamany Hall and constructed most of the major public works in New York City, including the Triborough Bridge, all of its expressways and parkways (a term he invented). This is a lengthy book, but it provides an interesting history of how cities are shaped.

Olmsted, Frederick Laws. ***Civilizing American Cities: Writings on City Landscapes.*** **Edited by S.B. Sutton. Cambridge, MA: Da Capo Press, 1997.**
Olmsted is best known for designing New York's Central Park, but his influence on the landscape of America is extraordinary. His designs brought trees and greenery to many urban spaces. This is a collection of Olmsted's plans for New York, San Francisco, Buffalo, Montreal, Chicago, Boston, and Berkeley, California, in addition to general writings on urban landscapes.

1. From "Intelligent Husbandry," in *A Documentary History of Conservation in America*, eds. Robert McHenry and Charles Van Doren (New York: Praeger Publishers, 1972), 276.
2. Charles T. Rubin, *The Green Crusade: Rethinking the Roots of Environmentalism* (New York: The Free Press, 1994), 6.
3. National Research Council, *Our Common Journey* (Washington, DC: National Academy Press, 1999), 93.
4. Robert M. May, "How Many Species are There on Earth?" *Science* 241 (September 10, 1998): 1441-1449, 1445.
5. J. A. McNeeley, et al., *Conserving the World's Biological Diversity* (Gland, SW: International Union of Conservation of Nature and Natural Resources, 1990).
6. Andrew Goudie, *The Human Impact on the Environment* (Cambridge, MA: MIT Press, 1996), 384.
7. It is interesting that the name "greenhouse effect," which was coined about a century ago, is scientifically inaccurate. A greenhouse stays warm because its closed windows prevent the inside air from giving its heat up to the outside air by convection, not because they block heat radiation.
8. British Petroleum, *1998 Statistical Review of World Energy* (Available at *http://www.bp.com*).
9. U.S. Department of Transportation, Bureau of Transportation Statistics, *North American Transportation Highlights* (December 1999) (Available at *http://www.bts.gov/*).
10. Environmental Protection Agency, *Environmental Fact Sheet: Partnership for a New Generation of Vehicles and the Environment*, No. EPA420-F-98-007 (April 1998).
11. Energy Information Agency, Department of Energy, *Describing Current and Potential Markets for Alternative Fuel Vehicles* (Available at *ftp://ftp.eia.doe.gov/pub/pdf/consumption/0604.pdf*).
12. S. S. Penner, "United States Energy Supplies for the 21st Century." *Energy* 23, no. 2 (1998): 71-78.
13. Energy Information Agency, *Electric Power Overview*, 1998 (Available at *http://w.eia.doe.gov/cneaf/electricity/pge/prim2/toc2.html*).
14. *Nuclear Notes From France* (November/December 1997) (Available at *http://www.info-france.org/America/embassy/nuclear/n2f2/november97*).
15. B. Magnus Francis, *Toxic Substances in the Environment* (New York: John Wiley & Sons, Inc., 1994), 279.
16. Kullervo Kuusela, *Forest Resources in Europe 1950-1990* (Cambridge, UK: Cambridge University Press, 1994).
17. National Research Council, *Our Common Journey* (Washington, DC: National Academy Press, 1999), 34.
18. *Our Common Journey*, 246.
19. Bruce Ames, Renae Magaw, and Lois Gold, "Ranking Possible Carcinogenic Hazards." *Science* 258 (April 17, 1987): 271-277.
20. National Research Council, *Hardrock Mining on Federal Lands* (Washington, DC: National Academy Press, 1999), Appendix B.
21. *Goudie*, 113.
22 Information from the Institute of Scrap Recycling Industries (*http://www.isri.org*).
23. World Health Organization, *The World Health Report 1999* (Geneva, SW: World Health Organization, 1999), 2.
24. The Population Reference Bureau's *1999 World Population Data Sheet* includes information on the availability of birth and death statistics by country. (Available at *http://www.prb.org/pubs/wpds99/wpds99_3.htm*).

25. Thomas R. Malthus, a British political economist, wrote a popular essay in 1798 predicting that, since population increases geometrically while food supplies increase arithmetically, eventually food supplies would become scarce.
26. In his best-selling 1968 book, *The Population Bomb*, Paul Ehrlich began by saying, "The battle to feed all of humanity is over. In the 1970s and 1980s hundreds of millions of people will starve to death in spite of any crash programs embarked upon now." He has written a more recent book on this topic, *The Population Explosion* (Simon & Shuster, 1990), with his wife, Anne Ehrlich.
27. United Nations Population Information Network, *World Population Prospects: The 1998 Revisions*. (Available at *http://www.undp.org/popin/popin.htm*).
28. U.N. AIDS Organization and World Health Organization, *HIV/AIDS: The Global Epidemic* (December 1996) (Available at *http://www.unaids.org/publications/epidemiology/estimates/situat96.kme.html*).
29. D. T. Allen and R. Jain, ed., "Special Issue on Industrial Waste Generation and Management." *Hazardous Waste and Hazardous Materials* 9 (1992): 1-11.
30. Thomas E. Graedel and Braden R. Allenby, *Industrial Ecology and the Automobile* (Upper Saddle River, NJ: Prentice-Hall, Inc., 1998), 147-148.
31. Paul Slovic, "Perceptions of Risk." *Science* 236 (April 17, 1997): 280-285.
32. Thalidomide has been licensed for use again in the U.S. as a drug therapy for AIDS-related conditions. For more on thalidomide, see Policy.com's coverage of this issue at *http://www.policy.com/issuewk/98/0817/081798l.html*
33. R. Neil Sampson, "Land Use," in *The Encyclopedia of the Environment*, ed. Ruth Eblen and William R. Eblen (New York: Houghton Mifflin Company, 1994), 405-408, 408.

Index